KU-731-129

WITHDRAWN

PERGAMON INTERNATIONAL LIBRARY
of Science, Technology, Engineering and Social Studies
The 1000-volume original paperback library in aid of education,
industrial training and the enjoyment of leisure
Publisher: Robert Maxwell, M.C.

TK
9145,
HUN

Fission, Fusion and the
Energy Crisis

SECOND EDITION

WITHDRAWN

THE PERGAMON TEXTBOOK
INSPECTION COPY SERVICE

An inspection copy of any book published in the Pergamon International Library will gladly be sent to academic staff without obligation for their consideration for course adoption or recommendation. Copies may be retained for a period of 60 days from receipt and returned if not suitable. When a particular title is adopted or recommended for adoption for class use and the recommendation results in a sale of 12 or more copies, the inspection copy may be retained with our compliments. The Publishers will be pleased to receive suggestions for revised edition and new titles to be published in this important International Library.

Other Pergamon Titles of Interest

Pergamon Related Journals

18 FEB 1980

Fission, Fusion and the Energy Crisis

S. E. HUNT

Ph.D., F.Inst.P., F.I.Nuc.E.
Member of the Electricity Supply Research Council
Professor of Physics at the University
of Aston in Birmingham, England

SECOND EDITION

PERGAMON PRESS

Oxford · New York · Toronto · Sydney · Paris · Frankfurt

UK	Pergamon Press Ltd., Headington Hill Hall, Oxford OX3 0BW, England
USA	Pergamon Press Inc., Maxwell House, Fairview Park, Elmsford, New York 10523, USA
CANADA	Pergamon of Canada, Suite 104, 150 Consumers Road, Willowdale, Ontario M2J 1P9, Canada
AUSTRALIA	Pergamon Press (Aust.) Pty. Ltd., P.O. Box 544, Potts Point, NSW 2011, Australia
FRANCE	Pergamon Press SARL, 24 rue des Ecoles, 75240 Paris, Cedex 05, France
FEDERAL REPUBLIC OF GERMANY	Pergamon Press GmbH, 6242 Kronberg-Taunus, Pferdstrasse 1, Federal Republic of Germany

Copyright © 1980 S. E. Hunt

All Rights Reserved. No part of this publication may be reproduced, stored in a retrieval system or transmitted in any form or by any means: electronic, electrostatic, magnetic tape, mechanical, photocopying, recording or otherwise, without permission in writing from the publishers

First edition 1973
Second edition 1980

British Library Cataloguing in Publication Data
Hunt, Stanley Ernest
Fission, fusion and the energy crisis. − 2nd ed.
1. Atomic power
I. Title
621.48   TK9145   79-40529

ISBN 0-08-024734-2 (Hardcover)
ISBN 0-08-024733-4 (Flexicover)

Printed and bound in Great Britain by William Clowes (Beccles) Limited Beccles and London

Contents

Preface to the Second Edition

SINCE the appearance of the first edition in 1974 the importance of the breeder reactor to the efficient use of the nuclear fuel reserves has become much more widely recognised, and with it the debate on the possible danger of the 'plutonium economy', briefly anticipated in the first edition, has become widespread. The second edition has been enlarged to examine these issues in more detail.

Despite the growing realisation that fissile fuel reserves are extremely limited in the absence of a substantial breeder-reactor programme, many countries are putting forward ambitious programmes based mainly on slightly enriched 'burner' reactors. The chapter on national programmes has been up-dated to include this, and the possible consequences of the policy, previously discussed in an appendix in later printings of the first edition, are now examined in the main text of the book.

Since 1974 some progress has been made in the development of renewable energy sources, notably solar energy and wave power, and these are discussed in more detail than previously, as are recent developments in the nuclear fusion research leading to the decision to site the JET project in the United Kingdom fusion laboratories at Culham. In this context reports from Princeton on a similar apparatus, P.L.T., are encouraging.

Whilst the potential value of these new energy sources should not be under-estimated, the global choice for the twenty-first century appears to remain between the development of the fast-breeder reactor, a more speedy exploitation of fossil fuel reserves or a 'levelling down' of *per capita* fuel consumption and with it general living standards.

In the author's view the latter possibility is not realistic in our present highly competitive society, and the second, if feasible, would represent a greater environmental and social hazard than would the development of the fast breeder. The recent American ban on fuel reprocessing and their rejection of the fast breeder is, therefore, particularly disturbing.

Preface to the First Edition

IT IS now some twenty years since nuclear power was hailed as the panacea of most, if not all, of the material problems besetting mankind. Present lay opinion is that this early promise has not been fulfilled, and it would be pointless to argue that the development of nuclear power has not experienced its fair share of difficulty and disappointment. There have, however, also been solid achievements and on occasions spectacular advances in this field over the past twenty years, and it seems appropriate to review the situation at this stage, when nuclear power is finally becoming competitive on a strictly economic basis and the availability of other fuels in the long term is rightly becoming a cause for concern.

Recent politically motivated restrictions in the supply of these fuels merely introduce in the mid-seventies a situation which was inevitable by the mid-eighties. Almost all the developed and some of the developing countries are now planning very significant expansions in their nuclear-power programmes, and the relevant question is no longer whether to build nuclear-power stations, but which type to choose from the many competitive designs available.

There is a clear conflict between the short-term exploitation and long-term development of nuclear power which should influence the choice of fission-reactor type, and an estimate of the probable outcome of the fusion-research programme is a very significant factor in this context.

This book is intended, to a large extent, for the undergraduate student and seeks to emphasise the inter-relationships of the scientific, technological, economic and ecological aspects of nuclear power. In the opinion of the author, so strong are these relationships that it is misleading, if not impossible, to study one of these factors without appreciable reference to the others.

The approach is descriptive, and by including the relevant basic atomic and nuclear physics as introductory material it is hoped that this book will also serve as an understandable survey of the present nuclear-power situation for the non-scientist. It is one of the author's more optimistic aims that it should be instrumental in persuading the arts graduate to take an increasing interest in this field, which has profound social as well as scientific importance. One must agree that the social consequences of technological development must be the subject of careful examination, but this cannot be done without some appreciation of the basic principles of the technology itself.

With the lay reader in mind, the use of technical jargon has been reduced to a minimum, but where it is necessary to introduce technical terms this is done with suitable explanation, and a Glossary of Terms is provided.

In the interests of clarity and brevity, the temptation to quote references in support of established scientific fact has been resisted. The factual material is widely accepted by reactor scientists. The interpretation and forward projections may be slightly more controversial, but the author feels them to be strongly supported by the evidence at present available.

Some ideas have been stolen from other people and in these cases sources are quoted wherever possible.

Acknowledgements to the Second Edition

THE author would like to express his renewed thanks to Mrs. I. Neal (*née* Templeton) for typing the extensive modifications incorporated in the second edition and to Mr. G. Smith for numerous modifications to the figures.

Thanks are also due to Dr. P. N. Cooper for his permission to use several illustrations.

Acknowledgements to the First Edition

THE author is indebted to Mrs. R. Finch for her care and occasional inspiration in converting his manuscript into a legible form, to Miss I. Templeton for her invaluable secretarial assistance and also to Miss F. A. Schofield and Mr. G. Smith for their work in producing the figures.

He would also like to thank colleagues at Aston and elsewhere for many useful discussions and in particular Professor J. Fremlin, Dr. P. N. Cooper, Dr. A. J. Cox and Mr. G. Williams for helpful comments on the manuscript.

Finally he would like to thank his wife and daughters for their assistance and tolerance throughout the rather prolonged gestation period of this book.

'THE ACCIDENT RECORD THAT PROVES NUCLEAR POWER IS SAFER'

Letter published in The *Guardian*, 11th April, 1979

Sir,

Harrisburg

The Pressurised Water Reactor (P.W.R.) is generally regarded as presenting a greater hazard than other reactor types because of its higher heat output per unit volume. Some seven per cent of this heat is due to the activity of the fission products and cannot be 'switched off' by shutting the reactor down. Consequently the danger of core melting when the coolant is lost due to mechanical fault or human error, or a combination of the two which appears to have been the case at Harrisburg, is greater for P.W.R.'s. than other thermal reactors. It is largely for this reason that the United Kingdom has so far opted in favour of its own gas cooled systems Magnox followed by A.G.R., and rightly so in my view.

Yet it must be appreciated that over 50,000 megawatts of P.W.R. and related B.W.R. stations are at present operating in the United States, and that these reactor types have types have been widely adopted throughout the world. So far there has not been a single fatality due to reactor malfunction, a safety record which is far superior to that of other methods of electricity generation. The reason lies in the extreme care in design and duplication of safety mechanisms which is a feature of all acceptable reactor design and strangely lacking elsewhere. At Harrisburg, for example, major mistakes appear to have been made and main components proved to be faulty yet these did not, in the event, result in a major release of activity.

In contrast three weeks ago, in the U.K., ten miners were killed reportedly due to a faulty air circulation system, and the annual casualties due to underground mining accidents amount to some sixty per year in this country alone. Disablement due to pneumoconiosis far exceeds this and if one is looking for an unquantified but major hazard, death of the general population due to 'bronchitis' accentuated by sulphur dioxide and other fumes from conventional coal and oil power stations far exceeds any which are likely to occcur from the by product of nuclear stations, including the long term disposal of radioactive waste by present and planned processes.

Sympathetic as one may be to the conservationist view, it is clear that the maintenance of our present living standards and the legitimate aspirations of developing nations for improved ones, depends critically on the availability of energy sources, with nuclear power and coal as the only proven long term alternatives. All objective assessments of the comparative hazards involved, when the full fuel cycle is considered in each case, are markedly in favour of nuclear power.

Yours faithfully,

PROFESSOR S. E. HUNT
Head of Department of Physics,
The University of Aston in Birmingham

CHAPTER 1

Living on Capital

IN FINANCIAL circles and even in the management of our personal affairs, to live from accumulated capital reserves is not regarded as a very satisfactory procedure, yet we are doing exactly this in the consumption of fuel to produce electricity and other forms of power on which our present standard of living is critically dependent.

In order to discuss the power requirements and available fuel supplies on a global scale, we must define a suitable unit. Economists have chosen the unit 'Q', which is equivalent to the energy produced by burning 46,500 million tons of coal, producing approximately 300 billion (3×10^{14}) kilowatt hours of electrical energy. This is, of course, an extremely large unit, but an appropriate one for estimating our present and future requirements. Perhaps the most significant factor to be considered is the rate at which these requirements have grown throughout our history. It is estimated that from the birth of Christ to the beginning of the Industrial Revolution, about A.D. 1850, the world consumption of energy was approximately 4 Q. This was principally produced by burning some of the current year's growth of wood and animal refuse, and during this time little, if any, use was made of the 'capital resources' of coal, oil, natural gas, etc., the so-called fossil fuels which had been accumulated since prehistoric times due to under-exploitation of growing fuel sources. These capital reserves were in fact continuing to accumulate during this period. By 1850 we were using fuel at the rate of some 1 Q per century, and a hundred years later, in 1950, the consumption had increased to approximately 10 Q per century. In 1970 alone we used approximately 0.2 Q, that is the rate was approximately double that in 1950, and immediately before the oil crisis the world energy demand was increasing with a ten-year doubling time. Forward predictions of fuel requirements are notoriously difficult; the main factors affecting these are clearly the growth of world population and the improvement in the standard of living, since, in the utilisation of conventional fuels at least, we can no longer reasonably expect any dramatic improvement in the efficiency at which fuel is converted into energy.

Per capita fuel consumption is very closely linked with the material standard of living. In round terms the average income in the United States is twice that in Western Europe, which, in turn, is some fifteen times greater than that in India, and the *per capita* fuel consumptions vary in almost exactly the same ratios. A similar relationship exists for countries between these limits, as shown in Fig. 1. It is perhaps pointless to argue whether the ready availability of fuel has led to a high standard of living or whether the desire for a high standard of living itself has generated the need for and development of high fuel consumption. This is a traditional 'hen and egg' situation. The relevant fact is that fuel consumption and standard of living go hand in hand, so that in predicting our future fuel requirements, some estimates of future standards of living are clearly called for. In the 'developed' countries such as the United States, Canada, Western Europe, etc., the average standard of living is still improving and the *per capita* fuel consumption appears to increase relentlessly at the rate of about 3% per year. Such is the economic balance that increases of slightly less than this figure coincide with periods of economic depression, slightly greater increases with periods of economic boom. Even in developed countries there does not appear to be as yet any saturation point at which power and fuel consumption may be expected to level off at a figure corresponding to some finally acceptable standard of living. For future planning, one possible postulate is that the world demand for fuel will increase continuously at a rate of 3% per head per annum as in the developed countries. When one considers that most of the large centres of population need to increase their national income and *per capita* fuel consumption by at least a factor of 10 before they reach a level of material comfort which we now regard as barely acceptable, it is clear that, for social and ethical reasons, there is strong argument that the world *per capita* fuel consumption should increase at a rate far in excess of 3% per annum even if it were to be frozen at present levels in developed countries.

Predictions of the growth of world population must also be uncertain. The early assumption that it would increase by a factor of 3 between 1950 and 2050 appears, in 1978, to be partially justified. The world population doubled in the previous century (1850-1950) in spite of two world wars and the relatively limited application of medical science.

Based on the above possibly conservative predictions, the world energy requirement between now and A.D. 2050 is estimated to be about 70 Q. Since this is many times greater than the world's energy consumption to date, it is relevant to examine our energy resources in some detail. As has already been indicated, these may be divided into two categories, the renewable sources, that is the energy which we receive each year, and the capital reserves which have accumulated in previous centuries when supplies exceeded demand.

Appreciable use was made of renewable energy sources before the Industrial Revolution, when wind and water power, albeit in a rudimentary form, were widely used and wood was the main fuel. With the Industrial Revolution coal became the main energy source, to be supplanted to a greater or lesser extent in many countries by 'cheap oil' from the Middle East in the post-war era.

The dramatic increase in oil prices in 1974 forced some reappraisal of this situation but the fundamental problem is not one of price but of overall reserves. The estimated global oil reserves are comparatively small, probably less than 2 Q, and despite the importance of the new North Sea fields to the British economy the rate of discovery of new oil reserves on a global basis has been falling steadily over the past twenty years whilst oil consumption still continues to grow despite the economies in its use by some industrial nations. The steady fall in the ratio of new discoveries to current consumption leads to an estimated peak in oil consumption in the early 1980s followed by a decline dictated by the exhaustion of the overall oil reserves.

In the early 1980s it is anticipated that approximately 55% of the total oil production will be in the Middle East, 20% in the United States, perhaps 7% from the North Sea fields and 18% from the rest of the world. This excludes oil production from the Communist countries. The oil shortage in the late eighties could be quite dramatic unless much more severe measures are taken to restrict its use as a fuel to those applications in which it has a clear advantage over other energy sources. Despite its comparatively high domestic production the United States is already an oil-importing nation and a realistic extrapolation of the American current consumption trends indicate that they would be capable of consuming the total free world oil production by the year 2020 (Fig. 2). It is clear that a move from oil as a main energy source is both desirable and urgent. Almost paradoxically whilst large quantities of oil are still being used to power electricity generating plant many countries are devoting considerable research effort into the problem of converting coal into a liquid fuel. This is intended to replace petrol in the specialised field of motor transport as the oil reserves become exhausted.

The coal reserves, viewed on a global basis, are in an appreciably healthier state than are the oil reserves. World reserves of mineable coal are estimated at about 8700 Gigatons* (or, say, about 190 Q), but as shown in Table 1 only Western Germany and the United Kingdom amongst the industrial countries of Western Europe hold significant reserves. The major coal reserves are in the U.S.S.R., U.S.A. and China which are remote from the Western European countries and many of the developing countries, so that the cost of transporting fuel, whether expressed in financial or energy terms, would be excessive. To transport coal, for example, from Siberia, where many of the Russian reserves are situated, to Western Europe would require a railway or other transport system far in excess of the existing one, and an appreciable fraction of the cargo would be consumed in transit. Reliance on coal as a long-term energy source could clearly pose appreciable economic and practical problems, even if political factors are ignored.

The so-called 'renewable' energy sources would in principle avoid the large-scale depletion of our capital fuel reserves. These involve harnessing the vast quantity of solar energy, about 5000 Q per year falling on the earth's surface, either directly or via its indirect manifestations such as wave and wind power, and will be discussed in greater detail in Chapter 11. Attractive as they are in principle it seems unlikely that the renewable energy sources can be developed to meet a

---

*1 Gigaton is equal to 1000 million tons.

TABLE 1. *Estimated World Fuel Consumption and Fossil Fuel Reserves*

Fuel consumption	
0 – A.D. 1850	4 Q*
About 1850	1 Q/century
About 1950	10 Q/century
About 1970	20 Q/century
2000 – 2050 (est.)	60 Q

Fossil fuel reserves		
Country	Reserves (Gigatons)	Q
Coal		
Australia	110	2.36
Brazil	11	0.24
Canada	85	1.82
China	1010	21.80
Colombia	13	0.28
Czechoslovakia	22	0.47
East Germany	30	0.65
West Germany	70	1.52
India	106	2.28
Japan	19	0.42
Poland	61	1.33
South Africa	72	1.55
U.S.S.R.	5527	119.00
United Kingdom	16	0.34
U.S.A.	1506	32.40
Yugoslavia	27	0.58
Total World	8685	187.04 Q
Non-Communist World		43.21 Q

Oil
About 2 Q, mainly in the Middle East.

* $Q \equiv 46{,}500$ million tons (46.5 Gigatons) of coal equivalent.

large fraction of the demand of an advanced civilisation. We must therefore continue to exploit our capital reserves, but this must be done with a greater regard to the long-term energy situation rather than just their immediate economically competitive position. This is particularly true of the other large capital energy reserve, fissile material. Here the situation is more complicated and, as explained in Chapter 7, rapid and uncontrolled exploitation could reduce the overall fuel value of our fissile reserves less than that of oil, whereas a more rational long-term programme, based ultimately on breeder reactors, could increase the fuel value of our fissile reserves to many times that of the coal reserves.

Nuclear fusion has been widely hailed as the panacea to our long-term energy problem. This involves the utilisation of our capital reserves of light elements as fuel, and the possible utilisation of tremendous reserves of deuterium in sea water have been widely publicised in the press. Progress in this exciting field will be reviewed in Chapter 12, from which it is clear that even if the necessary fusion conditions are achieved the residual engineering problems are considerable. Foreseeable generations of fusion reactors will be based on the deuterium-tritium reaction rather than the all-deuterium reaction, so that it is the much smaller capital reserves of lithium, from which tritium is produced, which will become the limiting factor.

Even a successful outcome of the fusion programme will not therefore avoid the necessity of a conservationist approach to fuel reserves of all types, just as with other raw materials on which society is increasingly dependent.

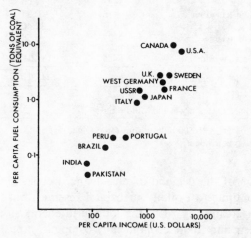

FIG. 1. Income vs. fuel consumption (mainly 1972 figures).

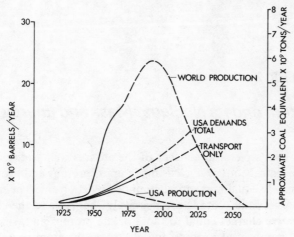

Fig. 2. Oil production and demand.

CHAPTER 2

The Atom and Its Nucleus, Mass and Energy

RUTHERFORD's pictorial concept of the atom is well known. It consists of a central core or nucleus made up of protons and neutrons in approximately equal numbers surrounded by electrons. The neutrons are uncharged, the protons are positively charged and the electrons have negative charges equal to those of the protons and, since there are equal numbers of protons and electrons, the atom as a whole is electrically neutral.

The electrons are attracted to the nucleus by electrostatic or coulomb forces, the magnitudes of which are proportional to the charges on the electron and the nucleus multiplied together, and divided by the square of the distance between them. Electrostatic forces of this type are well known and exist in classical 'large-scale' physics. Gravitational forces resemble them in that they are proportional to the product of the masses of the bodies involved and fall off with distance in the same way. It is not therefore surprising to find that the electrons orbit around the nucleus describing circular and elliptical paths in much the same way as the planets orbit around the sun, since the forces acting obey similar laws.

The radius of the outer electron orbit is, however, only about $1/10^{10}$ m (one-tenth of a thousandth of a millionth of a metre) and this may be taken as the radius of the atom. In dealing with systems of these dimensions, Bohr postulated that the electron orbits must be such that the electron energies must correspond to a finite number of small energy 'packets', a continuous distribution of electron energies not being permissible. The same is true for large-scale planetary motion, but the total energy involved is so large and is made up of so many energy packets that energy is effectively continuous and the concept loses its significance. It remains significant, however, in the electron case, and by applying this concept together with relatively simple mathematics it is possible to account for the motion of the electrons around the nucleus in some detail. By introducing the additional idea that no two electrons in an atom can be identical, it is possible to deduce a large amount of information about the chemical properties of the various elements, since these depend entirely on the configurations of the electrons, particularly the outer ones.

A somewhat more rigorous explanation for the behaviour of the electrons may be obtained by considering them as waves rather than particles. The results of this so-called wave mechanical approach are in general agreement with those obtained from the Bohr theory, and the minor differences are not relevant to the present discussion.

The mass of the electron is about $1/10^{30}$ kg (one-millionth of a billionth of a billionth of a kilogram) and the protons and neutrons are 1836 and 1838 times heavier than the electrons respectively. The radius of the nucleus is much smaller than that of the electron orbits, varying from about $1/10^{15}$ m (one-thousandth of a billionth of a metre) for the lightest nuclei to ten times this for the heaviest. From this it is seen that almost all the mass of the atom is concentrated in the nucleus which occupies only about one-billionth $(1/10^{12})$ of the total atomic volume. The density of the nucleus is therefore about one billion times greater than the density of the atom as a whole and therefore of solids and liquids which we meet in everyday life.

Since the nucleus consists of positively charged protons and neutrons, the electrostatic or coulomb forces between the particles of like charge (protons) are repulsive and tend to cause the nucleus to disintegrate. It would do so if it were not for another type of force, the so-called inter-nucleon or nuclear force, which is attractive and very much stronger than the electrostatic forces.

This force is not completely understood and does not appear to have a counterpart in other branches of physics, but we do know that it is of very short range, about $1/10^{15}$ m, or comparable with the nuclear dimension, and that within this range it is very strong indeed, approximately a million times stronger than the coulomb forces which bind the outer electrons to the nucleus. This fact is of great importance when we compare the energy release as a result of chemical reactions and nuclear reactions.

It should be stated initially that all reactions, whether chemical or nuclear, obey Einstein's mass-energy relationship, $E = m_0c^2$, where E is the energy released or absorbed in a reaction, m_0 the change in mass and c the velocity of light.

If we take as an example the simple chemical reaction of burning, which we can simplify as the combination of carbon with oxygen to form carbon dioxide (in chemical symbols $C + O_2 = CO_2$), careful weighing of the carbon dioxide formed would, in principle, show it to be slightly lighter than the combined weights of oxygen and carbon. The amount of mass missing appears in the form of energy (thermal energy or heat) according to Einstein's equation. However, in this reaction, as in all chemical reactions, the fraction of mass converted into heat is very small, about one part in a thousand million ($1/10^9$). This is because in chemical reactions we are merely rearranging the outer electrons of the atom and consequently only the relatively weak coulomb forces are involved.

Since a mass discrepancy of only one part in a thousand million is involved, early experimentalists may be excused for failing to notice it in normal chemical reactions. By and large, experimental results are not measured to this order of accuracy. By the same token it appears to be extremely wasteful to convert mass into heat energy by using the chemical or burning process. Last year we converted approximately 10,000 million tons of fossil fuels (coal, oil and natural gas) into useless ash and toxic fumes in order to produce heat with an intrinsic efficiency of one part in a thousand million. This is the more worrying when one realises that these fossil fuels are in many cases valuable raw materials for the chemical industries of the future.

It is interesting to note that primitive man in his use of fire to heat his cave was applying Einstein's mass-energy equivalence. This is perhaps the most outstanding example of the practical application of a physical principle predating its theoretical explanation. We have been applying the principle ever since with only marginal improvements in overall efficiency, since the upper limit of about $1/10^9$ is in fact fundamental for chemical processes.

Nuclear reactions involve changes in the configurations of the neutrons and protons making up the nucleus, and we have been able to induce such reactions only in recent years. Following Rutherford's earlier work with naturally radioactive material, the first 'man-made' nuclear reaction was produced by Cockcroft and Walton in 1932. By bombarding lithium with protons from an accelerator they were able to make some lithium nuclei capture an incident proton to form beryllium nuclei which in turn split up into two helium nuclei or α-particles. The reaction can be expressed symbolically as follows:

$$^7_3\text{Li} + {}^1_1\text{H} \rightarrow {}^8_4\text{Be} \rightarrow 2{}^4_2\text{He}$$

and is illustrated in Fig. 4. Here the usual chemical symbols are used, but a superscript is added to denote the mass number or total number of nucleons (neutrons plus protons) and a subscript to denote atomic number or number of protons. The equation therefore indicates that a lithium nucleus consisting of seven nucleons of which three are protons (and therefore four neutrons) captures a hydrogen nucleus or proton to produce a beryllium nucleus of eight nucleons (four protons and by deduction four neutrons) and this then splits up into two helium nuclei (or α-particles) each consisting of two protons and two neutrons. The mass converted into energy as a result of this reaction is about 2% of the mass of a proton or neutron or 2.5 parts in a thousand of the total mass of the particles involved in the reaction.

The efficiency with which mass is converted into energy as a result of nuclear reactions is in general of the order of one part in a thousand. This may not appear to be a very high figure, but it is nevertheless a million times higher than the figure for chemical reactions, and this difference arises naturally from the fact that the forces involved in nuclear reactions are a million times stronger than those involved in chemical reactions.

We should perhaps expect that the nucleus, consisting as it does of matter a billion times denser than that encountered in everyday life, and held together by forces which are a million times stronger than those which are normally experienced, would behave in an unfamiliar manner. In fact it obeys the established physical laws very well indeed and the difference between nuclear behaviour and that of more familiar systems is one of scale only.

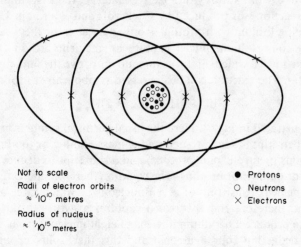

Not to scale
Radii of electron orbits
    $\approx \frac{1}{10^{10}}$ metres

Radius of nucleus
    $\approx \frac{1}{10^{15}}$ metres

● Protons
○ Neutrons
✕ Electrons

FIG. 3. The nuclear atom ($^{16}_{8}$O).

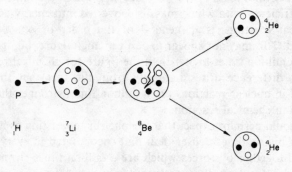

P

$^{1}_{1}$H

$^{7}_{3}$Li

$^{8}_{4}$Be

$^{4}_{2}$He

$^{4}_{2}$He

FIG. 4. The first 'splitting of the atom' ($^{7}_{3}$Li nucleus).

CHAPTER 3

Fission and the Bomb

As with so many major advances in science, the first experimentally observed fission process was the accidental by-product of a fundamental research programme aimed at quite a different objective. The heaviest naturally occurring nucleus is uranium which has two isotopes, $^{235}_{92}U$ and $^{238}_{92}U$. Since each nucleus has 92 protons it is surrounded by a shell structure containing 92 electrons and the atoms are therefore chemically identical. This is effectively the definition of the term isotope. The two isotopes have 143 and 146 neutrons respectively and since the nuclear configurations are different they would be expected to have different nuclear properties.

In 1938 Hahn and Strassmann were attempting to produce heavier nuclei than any occurring in nature by irradiating uranium with neutrons. On capturing an extra neutron, the neutron-rich uranium nucleus was expected to be unstable; one or more of its neutrons should change into a proton with the emission of an electron (a phenomenon known as β-decay first observed many years previously in naturally occurring radioactive nuclei). As a result of this process Hahn and Strassmann hoped to produce new nuclei with more than 92 protons and consequently more than 92 electrons in the electron shells. Such an atom would have chemical properties different from uranium or indeed from any known element. (Eleven such transuranic elements have since been artificially produced.)

Hahn and Strassmann observed that as a result of neutron irradiation their uranium specimen had indeed changed its chemical properties. In fact their 'transuranic' element appeared to have a bewildering range of chemical properties, many of which coincided not with those of the very heavy elements, but with the medium-weight elements. It soon became obvious that one of the neutron-rich uranium isotopes was unstable, not to the expected β-decay process, but to a much more violent interaction, the splitting of the nucleus into two slightly unequal parts with the emission of a small number of free neutrons. The isotope involved was $^{235}_{92}U$ and the reaction can be expressed symbolically as:

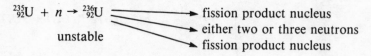

$$^{235}_{92}U + n \rightarrow {}^{236}_{92}U$$
unstable

fission product nucleus
either two or three neutrons
fission product nucleus

The unstable $^{236}_{92}U$ nuclei could break up in a wide variety of ways producing nuclei from germanium to gadolinium, but tend to break up rather asymmetrically to produce one nucleus containing about 140 nucleons and the other rather more than 90 nucleons (Fig. 5). Two features of this process made it of great practical interest. First it was exothermic, that is energy was given off, and, as in the case of most nuclear reactions, the fraction of mass converted into energy was about one part in a thousand. Secondly, the process produced free neutrons, usually either two or three, an average of about 2.5 per fission when one considered a large number of fission processes. Hence, we have an exothermic reaction which was produced by a neutron and which in turn produced neutrons. Such a reaction could, in principle, be self-sustaining if one could ensure that of these emitted neutrons at least one was captured by a $^{235}_{92}U$ nucleus to produce another fission. Such a reaction is known as a 'chain reaction'.

The date of the initial experiment, 1938, is significant, and the place, Berlin, even more so. Surprisingly, the potential military importance of the experiment, that it could produce an explosive approximately a million times more effective than any existing one, does not appear to have been fully appreciated in Germany. It was left to the Allies to exploit the discovery for military purposes, and their team contained a high proportion of scientists who had left Hitler's Germany for political or racial reasons. (A more detailed account of the history of this is given by Margaret Gowing (1964).)

It is not the purpose of this book to trace the development of nuclear weapons or to enter into the moral issues raised by their use. It is, however, useful to discuss the scientific principles on which they are based, as these have a relevance to the subsequent development of the nuclear-reactor programme.

It may appear a relatively easy task to ensure that from an average of 2.5 neutrons per fission, just one produces a fission in what we like to consider as the next generation of the process. Perhaps, fortunately, it is not. The reason is that the nucleus which undergoes fission (the fissile nucleus), $^{235}_{92}U$, occurs in natural uranium with an abundance of only 0.7%. The remaining 99.3% is made up of the heavier isotope $^{238}_{92}U$. On capturing a neutron, $^{238}_{92}U$ behaves exactly as Hahn and Strassmann expected it to do, that is it forms the heavier isotope $^{239}_{92}U$ which is unstable to decay by β-emission to form first neptunium and then plutonium. Neither of these elements occurs in nature. They are, in fact, the first two artificially produced transuranic elements. The reaction can be expressed symbolically as follows:

$$^{238}_{92}U + n \xrightarrow{\beta} {}^{239}_{92}U \xrightarrow{\beta} {}^{239}_{93}Np \rightarrow {}^{239}_{94}Pu.$$

As will be shown in Chapter 7, this reaction (known as the 'breeder reaction') is of utmost importance for the long-term future of the nuclear-reactor programme. If one is trying to produce a rapid self-sustained chain reaction using $^{235}_{92}U$ fission as the basis of a nuclear weapon it is, however, a nuisance. Most of the neutrons produced from the fission of $^{235}_{92}U$ are absorbed in the much more abundant $^{238}_{92}U$, so that not enough are available for capture in $^{235}_{92}U$ to sustain the fission chain. The initial approach in the 'Manhattan Project', as the Allied wartime nuclear-weapons programme was called, was to produce enriched $^{235}_{92}U$, which would support a sustained fission process, by separating it as far as possible from the $^{238}_{92}U$.

This could not be done chemically, since, as already mentioned, isotopes have identical chemical properties. More elaborate means of separation based on the fact that $^{235}_{92}U$ is slightly lighter than $^{238}_{92}U$ were therefore necessary. Because of this difference in mass, $^{235}_{92}U$ in the gaseous form of uranium hexafluoride ($^{235}_{92}UF_6$) diffuses through a suitable membrane approximately 1/3% faster than its heavier counterpart $^{238}_{92}UF_6$. By successive diffusion processes it is therefore possible to separate the two isotopes to a large extent, but it is clearly a tedious and expensive process. The uranium diffusion plant set up for this purpose at Oak Ridge, Tennessee, covered over a hundred acres. An alternative method of isotope separation was to ionise uranium, that is to remove one electron from the atom by an electrical discharge so that the resulting atom has a positive charge, and then to pass accelerated ions through a magnetic field. The ions describe circular paths and the radius of the $^{235}_{92}U$ ion trajectory is a fraction of a per cent smaller than that of the $^{238}_{92}U$ ion, so that the two may be collected separately. This method of separation is also slow and expensive, but between the two processes it was possible to accumulate sufficiently highly enriched $^{235}_{92}U$ to manufacture the first type of atomic bomb. Subsequently use was made of the breeder process to produce $^{239}_{94}Pu$, since this man-made nucleus also undergoes fission after neutron capture and, being chemically different from the uranium in which it is produced, can be separated relatively easily.

Even a highly enriched $^{235}_{92}U$ or $^{239}_{94}Pu$ assembly is not of necessity capable of sustaining a chain reaction, since an appreciable fraction of the neutrons produced may escape from the surface. In order to reduce the fraction of neutrons lost from the surface to an extent that a chain fission reaction can proceed, the assembly must have a certain minimum size, the so-called 'critical size'.

This can best be understood by reference to Fig. 6. This depicts two hemispheres of highly enriched $^{235}_{92}U$, A and B. Let us assume that each in itself is just 'subcritical', that is just too small to allow a chain reaction to be maintained because the fraction of neutrons lost, or 'leaked' at the surface, is too high. If now the two hemispheres are brought together, the previous leakage of neutrons across the two plane surfaces is eliminated; those neutrons previously lost from A across this surface now enter B and vice versa. The spherical configuration so formed will therefore support a sustained chain of fission reactions, that is it is above the critical size.

The rapid joining of slightly subcritical enriched $^{235}_{92}U$ assemblies was the basis of the detonation of the first type of fission nuclear weapon. Figure 7 illustrates the principle of a rapidly growing (or diverging) fission chain reaction produced in a highly enriched system of above critical size. The time between successive generations of fission processes is appreciably less than one-billionth of a second.

The later type of nuclear weapon, the so-called 'hydrogen bomb', relies for its explosive power on the fusing together of light nuclei, which, like the fission of some of the very heavy nuclei, is an exothermic process. The fusion process will be discussed in detail in Chapter 12, which deals with the attempts made to control the process in order to use it for the production of heat and electricity.

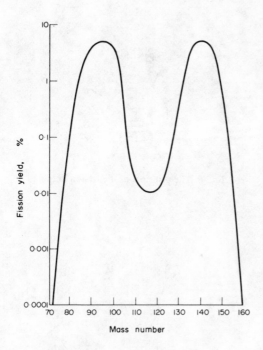

FIG. 5. Fission yield vs. Mass numbers.

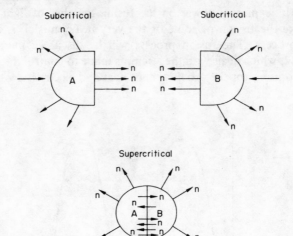

FIG. 6. The principle of the fission bomb.

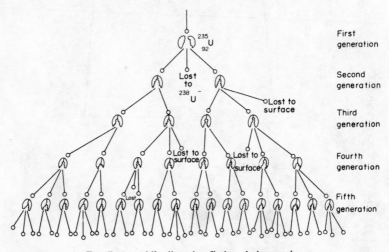

FIG. 7. A rapidly diverging fission chain reaction.

Reference

GOWING, MARGARET (1964) *Britain and Atomic Energy 1939-45*. Macmillan, London.

CHAPTER 4

The Natural Uranium Reactor

THE low isotopic abundance of fissile $^{235}_{92}$U makes it impossible to produce a sustained fission chain reaction in a simple block of natural uranium. The neutrons ejected at fission share the kinetic energy released by the fission process and are travelling at high velocities. $^{235}_{92}$U and the $^{238}_{92}$U nuclei have comparable capture probabilities or 'capture cross-sections' for fast neutrons, $^{235}_{92}$U capturing neutrons about six times more readily than $^{238}_{92}$U. Since there are about 140 $^{238}_{92}$U nuclei for each $^{235}_{92}$U nucleus, most of the fission neutrons would be captured by the $^{238}_{92}$U, and not enough captured by the $^{235}_{92}$U to sustain the chain reaction.

The situation is, however, different for very low-energy neutrons. These are captured by $^{235}_{92}$U to produce fission about 200 times more strongly than by $^{238}_{92}$U, and this more than compensates for the lower isotopic abundance of the fissile isotope $^{235}_{92}$U. A sustained chain of fission reactions is just possible if the fission neutrons can be slowed down from their initial high velocities to very low velocities, but the situation is complicated by the fact that neutrons with velocities intermediate between these two limits are captured by $^{238}_{92}$U even more strongly than by $^{235}_{92}$U. Figure 8 shows the energy distribution of neutrons immediately after fission and the way in which the capture probabilities (capture cross-sections) vary with neutron energy for $^{235}_{92}$U and $^{238}_{92}$U. (Note that the energy units used for fission neutrons are millions of electron volts and those used for the capture probability graphs are electron volts. These units are defined in the Glossary of Terms.)

The problem is to ensure that energetic fission neutrons are slowed down to very low energies without too many of them being caught in the so-called 'resonance trap' of $^{238}_{92}$U during the process (Fig. 8C). The slowing down, or moderating, process must be confined as far as possible to a region where there is no $^{238}_{92}U$ (nor of course any $^{235}_{92}U$, since the two are intimately mixed).

The neutrons are moderated by allowing them to collide with nuclei of the light elements. In so doing they transfer some of their energy to these nuclei in much the same way as a moving billiard ball does on striking a stationary one. It can easily be shown that for a 'head-on' collision the ratio of

$$\frac{\text{Neutron energy after impact}}{\text{Neutron energy before impact}} = \frac{(M_m - M_n)^2}{(M_m + M_n)^2}$$

where M_m and M_n are the masses of the moderating nucleus and neutron respectively.

If the moderating nucleus is that of the hydrogen atom (i.e. a proton), its mass M_m is for all practical purposes equal to that of the neutron M_n and the neutron can transfer all its energy to the proton in a single head-on collision. If the scattering nucleus contains twelve nucleons, $M_m = 12M_n$, the neutron can transfer only about 20% of its energy at one head-on collision, and as the moderating nuclei becomes heavier the fraction of energy lost by the neutrons for a single head-on collision is reduced. Only the lighter elements can therefore be used as moderators. In fact, very few of the actual collisions are 'head-on'. Usually the neutrons strike the moderating nuclei at an angle and the fraction of energy transferred per collision is on average very considerably less than for the head-on case. The average number of collisions required to slow the fast neutrons down to so-called 'thermal' energies, that is energies which are comparable to those due to thermal motion, for various light nuclei are as follows:

Element	Number of nucleons	Number of collisions
Hydrogen	1	18
Deuterium		
(heavy isotope of hydrogen)	2	25
Helium	4	43
Lithium	6 or 7	60
Beryllium	9	86
Boron	10 or 11	95
Carbon	12	114

By comparison, over 2000 collisions are required with uranium nuclei to produce this reduction in energy.

We also need to slow the neutrons down in as small a space as possible, so that apart from the requirement of maximum energy transfer per collision the moderator must be available in a reasonably concentrated form, that is as a liquid or a solid rather than a gas.

Of the substances listed, clearly hydrogen in the form of water, H_2O, appears, at first sight, to be ideal, and it is also readily available and cheap. Unfortunately it captures neutrons too readily, and if one uses natural uranium as fuel there are just not enough neutrons to spare.

The heavy isotope of hydrogen, deuterium, does not capture neutrons to any appreciable extent and may be obtained in a concentrated form as 'heavy water', D_2O. Although expensive, it is one of the two possible moderator materials for natural uranium reactors. Helium is chemically inert and can therefore only be obtained in its gaseous form, whilst lithium, beryllium and boron must be rejected on the grounds of chemical reactivity, toxicity, high capture cross-section or, so far as large natural uranium reactors are concerned, expense. This brings us to the second feasible moderating material, carbon, which is used in the form of graphite. Heavier nuclei produce insufficient energy transfer per collision.

The use of natural uranium, therefore, imposes two basic limitations in reactor design. The reactor must be heterogeneous, that is the fuel and moderator must be separated from each other as shown in Fig. 9 to minimise neutron capture by $^{238}_{92}U$ during moderation, and the moderator itself must be either heavy water or graphite. These restrictions are imposed because of the need to reduce the percentage of neutrons removed from the fission reaction chain to a minimum.

As is the case for nuclear weapons already discussed, it is necessary to minimise the fraction of neutrons escaping from the surface. The number of neutrons produced is proportional to the volume of the assembly and the number lost from the surface is proportional to the surface area. The fraction of neutrons lost at the surface is therefore proportional to the surface area/volume ratio of the assembly and this is reduced as the dimensions of the assembly are increased. This can easily be understood by considering a simple cubic configuration with sides of length x, then

$$\frac{\text{Number of neutrons escaping from surface}}{\text{Number of neutrons produced}} \propto \frac{6x^2}{x^3} \propto \frac{6}{x}.*$$

The fraction escaping is clearly reduced as x increases.

Since the fraction of neutrons which one can afford to lose from a natural uranium assembly is much smaller than that from an enriched system, the size necessary to ensure that the chain reaction can continue (the critical size) is much greater; a typical volume for a natural-uranium reactor is about 2000 m^3 compared to the order of 5 m^3 for a highly enriched system. Ideally the core should be spherical, since this shape has the minimum surface area/volume ratio, and hence the minimum neutron loss at the surface.

In the natural-uranium reactors it is also necessary to select the constructional materials used in the reactor with care, since any substance capturing neutrons strongly must be excluded. The situation is made more difficult by the fact that some of the nuclei produced by the fission process, so-called 'fission product nuclei', particularly xenon and samarium, themselves capture neutrons strongly. In designing a reactor it is therefore necessary to make allowances for this source of neutron loss (known as fission product poisoning) which develops during the operating life of the reactor. This point is considered in more detail in Chapter 6.

The sources of neutron loss in a natural-uranium reactor just above criticality are shown diagrammatically in Fig. 10. The time between one generation of fission processes and the next is about one-thousandth of a second, since this is the time required to moderate the neutrons by multiple collision. The number of fission processes in any given 'generation' of the chain process divided by the number of fissions in the previous generation is known as the multiplication factor of the system K. Clearly if K is greater than unity, the chain reaction will grow in magnitude or diverge and if K is less than unity it will decrease in magnitude or converge. In the former case we say that the system is 'supercritical' and in the latter case 'subcritical'. When the

---

*The symbol $\propto$ means 'proportional to'.

number of fissions does not change from one generation to the next the system is critical and a nuclear reactor in this condition will run at constant power. The chain reaction shown in Fig. 10 is very close to this. It will be seen that whilst some fission product nuclei absorb neutrons strongly, others emit neutrons some time after the fission process. These so-called 'delayed neutrons' make up about 0.65% of all the neutrons released in the fission processes. It will also be seen that a small fraction of the $^{235}_{92}U$ nuclei do not undergo fission as a result of neutron capture.

It is perhaps unfortunate for the image of nuclear power that the development of the natural-uranium reactor owes so much to the military programme. The first self-sustained fission chain reaction was produced in Chicago in 1942 and once the principle had been confirmed this was followed by the construction of larger reactors at Hanford. Each consisted of a heterogeneous system of natural-uranium fuel rods embedded in a block of graphite which acted as the moderator. The objective of the 'Hanford pile' was to produce man-made fission material $^{239}_{94}Pu$ by the 'breeder' reaction

$$^{238}_{92}U + n \rightarrow {}^{239}_{92}U \overset{\beta}{\rightarrow} {}^{239}_{93}Np \overset{\beta}{\rightarrow} {}^{239}_{94}Pu.$$

The $^{239}_{94}Pu$ was at that time required to produce the highly enriched supercritical system which was the basis of the second fission bomb. Even though the $^{239}_{94}Pu$ could be separated chemically from uranium, this involved the remote processing of the rods which were highly radio-active, due not so much to the plutonium as to the fission product nuclei which inevitably accumulate in used fuel. (The $^{238}_{92}U$ from which the fissile $^{239}_{94}Pu$ is produced is known as a fertile material.)

It is illustrative of the expense of the physical $^{235}_{92}U$ separation programme that it proved economically feasible to build what was effectively the first large-scale nuclear reactor, to produce plutonium and to carry out the difficult chemical separation process to obtain the alternative fissile material $^{239}_{94}Pu$.

Britain can justly claim to have pioneered the development of the natural-uranium graphite-moderated reactor for peaceful purposes. After extensive fundamental research and development work carried on mainly within the Atomic Energy Research Establishment at Harwell, the first significant power-producing reactor was constructed at Calder Hall in Cumberland, producing electrical power in 1956. The reactor core consists of vertical metallic natural-uranium rods or fuel elements forming a lattice within the graphite moderator. As previously indicated, the critical size of a natural-uranium reactor is large, and the Calder Hall core consisted of a 37-ft-diameter cylinder 27 ft high containing 120 tons of uranium fuel and 1150 tons of graphite moderator. The cylindrical shape was chosen since it was easier to construct than the ideal spherical assembly.

It should be appreciated that in a power station the nuclear reactor merely acts as a source of heat, replacing the furnace in the conventional power station. The heat produced in this reactor is extracted by blowing carbon dioxide at a pressure of 7 atm through the core. The carbon dioxide is then passed through heat exchangers to heat water and produce steam to operate the turbines and generators as for conventional power stations.

The reactor core is surrounded by a 2-in.-thick steel pressure vessel to contain the pressurised carbon dioxide and provide some shielding of personnel from the neutrons, γ-rays and other radiation which are produced with very high intensity at fission and subsequently by the fission product nuclei. The main radiation shield is an 8-ft.-thick concrete envelope which surrounds the reactor vessel. The power of the reactor is controlled by the use of vertical boron steel rods which absorb slow neutrons very strongly indeed and so compete with the fission reaction. By inserting the rods further into the core, the proportion of neutrons captured by $^{235}_{92}U$ is reduced and the reactivity* of the system and the power output consequently fall. The power may be increased by raising the rods. In normal operations the control rods are operated automatically to keep the power output at a preset level. The control and safety of reactors will be dealt with in more detail in Chapter 6.

The design power output of the Calder Hall reactor was 182 MW of heat, which was converted into 35 MW of electrical power, that is

---

*Reactivity is equal to the fraction by which the multiplication factor exceeds unity (reactivity $\Delta K = (K-1)/K$).

enough to operate 35,000 1-kW electrical heaters.* The actual uranium consumption is about ½ lb per day, less than one-fiftieth of the coal which would be required to produce just 1 kW of heat throughout this period. This reactor served as the prototype for the first generation of nuclear power stations in this country, the so-called 'Magnox' reactors, which get their name from the apparently minor design feature that the metallic uranium fuel is encapsulated or 'canned' in a protective container of magnesium alloy. This material was chosen because it does not readily capture neutrons. The further development of this reactor type will be discussed in Chapter 9, and a typical reactor in the series, Dungeness A, is shown in Fig. 11.

As already explained, the reactor is designed to reduce the fraction of neutrons captured by $^{238}_{92}U$ in order to achieve criticality. Consequently the amount of $^{239}_{94}Pu$ 'bred' as a result of this reaction is limited. The so-called 'conversion ratio', that is the amount of $^{239}_{94}Pu$ produced compared to the $^{235}_{92}U$ consumed for a natural uranium graphite system, is limited to about 0.8.

The only other type of reactor which is possible using natural uranium is a heterogeneous heavy-water-moderated system and the development of this type of reactor was pioneered by the Canadians in the so-called 'Candu' (Canada can do it) programme.

A pilot reactor of this type was the Demonstration Power Reactor which was constructed at Des Joachims and became critical in 1962. It consisted of clusters of natural-uranium fuel rods surrounded by heavy water under pressure, in horizontal tubes. The pressurised heavy water was circulated, transferring the heat produced by the fission processes to normal 'light' water via heat interchangers, and the subsequent electricity production was conventional. Some neutron moderation takes place in the pressurised heavy water, but the main moderation occurred in a large container (or calandria) of heavy water in which the pressurised tubes were immersed. The output of the Demonstration Power Reactor was 83.3 MW of heat converted to 19.3 MW(e) of electrical power. It was followed by a larger reactor of similar design at Douglas Point in Ontario (Fig. 12). This was the first

---

*The output of this reactor has now been increased to 258 MW of heat producing 50 MW of electricity.

really significant heavy-water-moderated power reactor with a heat output of 693 MW and an electricity output of 220 MW(e). It contains 416 tons of natural uranium and 1440 tons of heavy water moderated within a horizontal cylindrical stainless-steel vessel approximately 20 ft in diameter and 17 ft long. The conversion rate is almost 0.9.

Control of the heavy-water-moderated reactors is by cadmium rods since cadmium also captures slow neutrons strongly, but as an additional safety feature the heavy-water moderator can be emptied into a 'dump tank' situated underneath the reactor. When the moderator is removed in this manner the neutrons cease to be preferentially captured by $^{235}_{92}U$ and the fission chain reaction rapidly dies away. Like Calder Hall, and indeed all nuclear-power reactors, the Douglas Point reactor is surrounded by thick concrete radiation shielding.

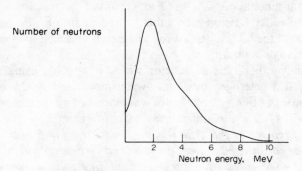

FIG. 8A. The fission spectrum of neutron energies.

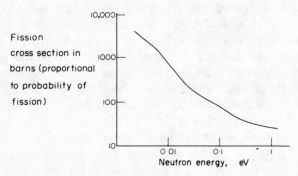

FIG. 8B. Fission probabilities (cross-sections) vs. neutron energy for $^{235}_{92}U$.

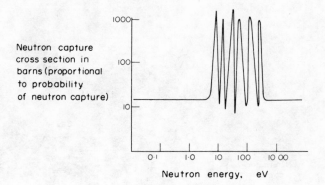

Neutron capture
cross section in
barns (proportional
to probability
of neutron capture)

Neutron energy, eV

FIG. 8C. Neutron capture probabilities (cross-sections) vs. neutron energy for $^{238}_{92}$U.

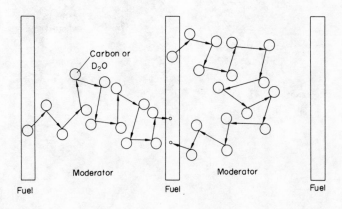

FIG. 9. Neutron moderation in a heterogeneous reactor.

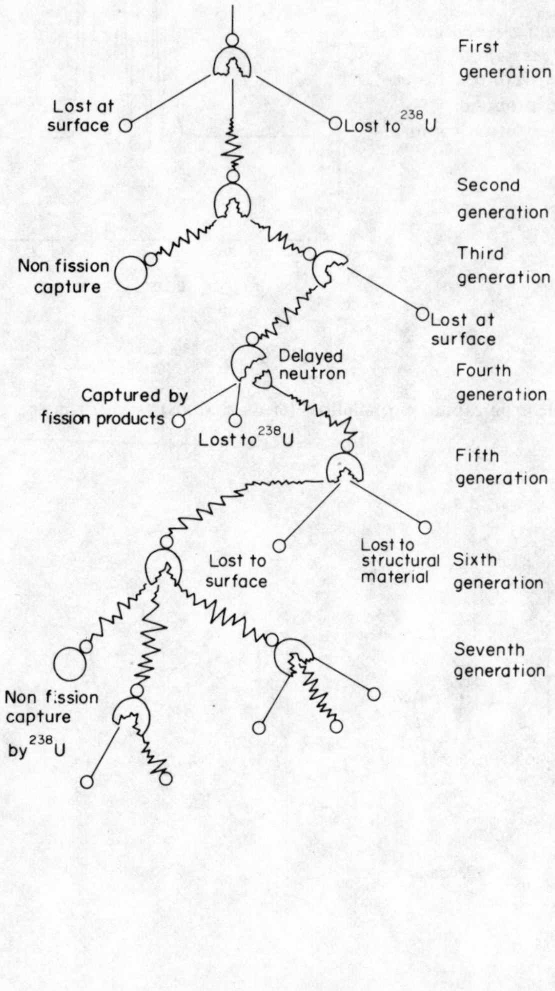

First
generation

Lost at
surface

Lost to ^{238}U

Second
generation

Non fission
capture

Third
generation

Lost at
surface

Captured by
fission products

Delayed
neutron

Fourth
generation

Lost to ^{238}U

Fifth
generation

Lost to
surface

Lost to
structural
material

Sixth
generation

Seventh
generation

Non fission
capture
by ^{238}U

FIG. 10. A chain reaction in a moderated natural-uranium assembly slightly
above criticality.

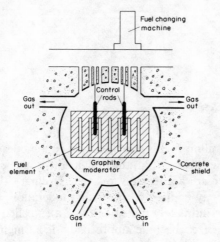

FIG. 11. A 'Magnox' reactor (Dungeness A).

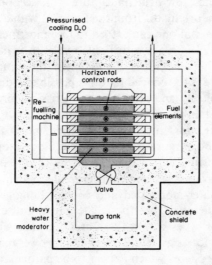

FIG. 12. A 'Candu' reactor (Douglas Point).

CHAPTER 5

Enriched Reactors

IF NATURAL uranium fuel is enriched even to a modest degree by the addition of fissile $^{235}_{92}U$ or $^{239}_{94}Pu$, the design constraints discussed in the previous chapter are appreciably relaxed. Ordinary (light) water or organic liquids may be used for neutron moderation, the moderator and fuel may be intimately mixed to form a so-called homogeneous system, critical sizes become appreciably less, high-melting-point alloys or uranium oxide rather than the metal itself may be used to increase the maximum operating temperature and constructional materials may be chosen primarily for their mechanical properties rather than for their low neutron capture cross-sections. This greater freedom in design originates from the fact that with a higher proportion of fissile material one can afford to lose more neutrons and still ensure that enough interact with fissile nuclei to maintain the chain reaction. At still higher degrees of enrichment the competitive capture by $^{238}_{92}U$ in the core is so reduced that it is possible to dispense with the moderator completely and rely mainly on fissions produced by fast neutrons to maintain the fission chain.

For enriched reactors the number of permutations and combinations of type and configuration of fuel and moderator becomes very large indeed. Existing and planned reactors can be divided roughly into four main groups: heterogenous graphite-moderated gas-cooled, heterogeneous water-moderated and cooled, homogeneous reactors and fast (unmoderated) reactors. The present discussion will be limited to reactor types within these broad groups which have been tried with some success as actual or potential power reactors, together with mention of systems which appear to be particularly attractive as future reactor types.

Perhaps the most conservative in principle of the enriched reactor designs is the Advanced Gas-cooled Reactor (A.G.R.) or Mark II reactor which is the basis of the second generation of nuclear-power reactors in this country. Like the Magnox or Mark I reactors, A.G.R.s are of heterogeneous design, graphite moderated and carbon dioxide cooled, but since the fuel contains about 2.5% of fissile material rather than 0.7%, it is possible to use high-melting-point uranium oxide rather than metallic uranium as fuel and this allows a considerable increase in the operating temperature. About 660°C is the present design aim. This is important, because the fraction of the heat which can be extracted from a reactor and converted into electricity increases as the gas-outlet temperature increases, and hence the overall efficiency increases with maximum operating temperature.

(Carnot cycle efficiency $= \dfrac{T_1 - T_2}{T_1}$ where T_1 and T_2 are the output and input temperatures on the absolute scale.)

For example, the efficiency with which heat from the Magnox reactors with output temperatures of 390°C is converted to electricity is about 30%; that for the A.G.R.s is over 40%.

The core of the reactor is much smaller, being about one-quarter of that for a Magnox reactor of the same power output. The problem of removing the heat from the fuel elements, where most of it is produced, is therefore more exacting than for the Magnox reactors, but fortunately, because of the reduced need for neutron economy, the fuel elements can be clad in more rugged stainless steel rather than magnesium alloy. The smaller fuel 'pins' are assembled in clusters and designed to present as large a cooling surface as possible to the pressurised carbon dioxide coolant. The heat interchanger is inside the pressure vessel (Fig. 13).

The A.G.R.s are, in fact, a logical development of the Magnox reactors. There is no appreciable conceptual change, but the reduced need for neutron economy has been used to improve the overall efficiency of the system, but unfortunately the conversion ratio of about 0.5 is less than that achieved in Magnox. A prototype A.G.R. has been in operation at Windscale in Cumberland since 1963 and the first commercial reactor of this type became operational in 1976 (Chapter 9).

A somewhat more adventurous development along similar general lines is the High Temperature Reactor (H.T.R.) developed at Winfrith (Dorset) as a collaborative European project involving France, Germany and the U.K., where an experimental 20-MW reactor 'Dragon' operated from 1965 to 1976. The fuel is slightly enriched uranium oxide in the form of small pellets. Each pellet is encased in a layer of carbides and carbon which not only serves as moderator but also contains the gaseous fission products which can cause swelling and mechanical failure in more conventional fuel rods. The pellets themselves are enclosed in a graphite matrix which provides additional moderation and also reflects escaping neutrons back into the fuel element. In order to minimise corrosion, the cooling gas is inert helium rather than carbon dioxide and the design maximum working temperature is above 1000°C. The conversion ratio of this type of reactor is higher than that of Magnox or A.G.R., being about 0.85, and its economically competitive position may depend to some extent on the direct operation of turbines by the high-temperature helium gas rather than using an intermediate steam circuit. The water-cooling requirement for the H.T.R. is less than that for most other types, which could reduce siting limitations. The mechanical design is similar to A.G.R.

A 'pebble-bed' version of the H.T.R. is being developed in Western Germany. In this type the enriched fuel is mixed with fertile thorium and encapsulated within carbides and carbon to form larger 'pebbles' which are fed into the top of the reactor, the spent fuel being extracted at the base. Figure 14 shows a schematic design.

Another main group of reactor designs use water rather than gas as the heat-transfer medium. Most of the development work on the heterogeneous light-water-moderated reactors was done in the United States. A fairly modest enrichment of between 2% and 5% is adequate to compensate for the increased neutron loss to the hydrogen of the light-water moderator and to allow reasonable freedom in the choice of constructional materials and design parameters generally. These reactors may be divided into two types: the boiling-water reactors (B.W.R.) in which the water, which serves as moderator and coolant, is actually allowed to boil inside the reactor vessel, and the pressurised-water reactors (P.W.R.), in which the pressure inside the reactor vessel is high enough to prevent boiling up to temperatures of about 350°C.

The P.W.R. operates at pressures some 160 times greater than normal atmospheric pressure. The high-pressure water from the reactor vessel is circulated through heat interchangers where it gives up most of its heat to a secondary water system at a lower pressure, so that boiling occurs in the secondary circuit and the steam is used to generate electricity in the normal manner. The prototype P.W.R. 'Yankee' was built in Massachusetts, U.S.A., and reached its full power of 600 MW(t) producing 172 MW of electricity in July 1961. It is illustrated diagrammatically in Fig. 15. The fuel was in the form of uranium rods enriched to 3.4% with ^{235}U, and these were grouped together to form a cylindrical core weighing some 21 tons. The reactor was controlled by rods which contained 80% silver, 15% boron and 5% cadmium, all of which elements absorb low-energy neutrons strongly. The reactor core and moderator were contained in a steel pressure vessel some 30 ft long and 9 ft in diameter, which had walls 8 in. thick in order to withstand the very high internal pressure. The conversion ratio of this reactor was about 0.55.

In the Boiling-Water reactors the pressure of the water is somewhat less, allowing boiling to occur in the water surrounding the reactor core.

Typically the water is maintained at a pressure of 1000 psi (about 70 atm), allowing boiling to occur at temperatures of about 300°C. The high-temperature high-pressure steam from the reactor core is then used directly to drive the turbines before being condensed and returned to the reactor vessel. The prototype B.W.R. 'Dresden' (Fig. 16) was constructed in Illinois, U.S.A., and came into operation in June 1960. The output was 680 MW of heat producing 200 MW(e) of electricity and the core consisted of 57.5 tons of uranium oxide enriched to 1.5% of ^{235}U. The reactor core and boiling water moderator were contained in a 5.5-in.-thick pressure vessel some 35 ft in diameter and 70 ft high. The conversion ratio was 0.6.

P.W.R. and B.W.R. are known as 'light-water reactors' (L.W.R.) as opposed to the heavy-water-moderated natural-uranium reactors of the Candu series. The one British reactor design using water moderation, the Steam Generating Heavy Water Reactor, is a hybrid of the

Canadian and American systems in that, like Candu, the fuel elements are contained in pressurised tubes but cooled by ordinary (light) water. Most of the neutron moderation takes place in heavy water at atmospheric pressure, the tubes containing the fuel elements and pressurised light water being immersed in the heavy-water container or 'calandria'. In this way the neutron economy possible by the use of heavy-water moderator is retained to a large extent, but the possible losses of heavy water are avoided by not using it as the heat-transfer medium under high pressure. The necessity for the large high-pressure vessel of the P.W.R or B.W.R. reactors is avoided, the comparatively narrow-bore tubes which contain the pressurised water being appreciably cheaper. A 100-MW(e) reactor of this type has been in operation at Winfrith since 1965 and is shown diagrammatically in Fig. 17.

In fact there are over 400 vertical pressure tubes containing slightly enriched uranium dioxide fuel elements each of which consists of 36 small 'pins', on the same general lines as the A.G.R. elements, clad in zirconium alloy. About 12% of the pressurised light water is converted to steam within the pressure tubes and is used to drive the turbines before condensation and recycling. The calandria contains about 120 tons of heavy water at atmospheric pressure. The reactor is controlled by raising or lowering the level of this heavy-water moderator and rapid emergency shutdown can be effected by injecting boric acid solution, which captures neutrons strongly, into the moderator.

Although in its present form the S.G.H.W.R. uses enriched fuel, it would be possible by increasing the D_2O/H_2O ratio to use natural uranium, an important factor in the long-term application of these reactors, which will be discussed in Chapter 7.

The reactor systems discussed so far, with the possible exception of H.T.R., are classed as heterogeneous systems, that is fuel and moderator are separate. The use of enriched fuels enables critical systems to be designed in which the fuel and moderator are intimately mixed, either in the form of a suspension or 'slurry' of fuel in a liquid moderator, fused salts of fuel and moderator, or a true solution of a uranium salt in water. A small reactor of the latter type was first constructed at Oak Ridge, Tennessee, in 1953.

The reactor consisted of a solution of highly enriched uranyl sulphate in water. The degree of enrichment was over 90% and because of this the critical size was very small. The whole 'reactor' was contained in a spherical stainless-steel container only 18 in. in diameter at a pressure of 1000 psi, and the solution, which in fact was the reactor core as well as the moderator, was circulated through a heat interchanger to raise steam in a secondary water circuit. The speed of circulation was about 10 l/sec and the miniature reactor had an output of 1 MW with an outlet temperature of about 250°C and an inlet temperature of about 200°C.

One attractive and novel feature of the reactor was that it was self-regulating, that is the heat produced adjusted itself automatically to meet the required load. If more heat was extracted from the heat interchanger the temperature of the returning liquid was reduced, it was consequently denser and acted as a more efficient moderator, so that the reactivity increased and with it the total power production. If the power requirement fell, less heat was extracted from the heat interchanger, the temperature of the returning liquid increased, its mean density and efficiency as a moderator fell and the reactivity of the system also fell to correspond to the reduced power demand. This system appears, superficially at least, to have many advantages.

In spite of these obviously desirable features, an early programme for the development of homogeneous reactors of this type (the homogeneous reactor experiment or H.R.E.) at the American Oak Ridge National Laboratories was disappointing due to corrosion problems and the complexity of the chemical processing plant required. They have been replaced by a type similar in principle but employing molten salts rather than solutions.

A small 8-MW reactor of this type has been in operation at Oak Ridge since late 1966. The core consists of a mixture of molten uranium, lithium, beryllium and zirconium fluorides which are not corrosive to the bare graphite rods, which provided additional moderation up to the opening temperature of 650°C.

As will be shown in Chapter 7, the long-term future of nuclear power depends vitally on the development of reactor systems of high conversion ratio, and conversion ratios in excess of unity cannot be achieved for the uranium-plutonium cycle in reactors using moderators, the so-called 'thermal reactors'. This is because of the build up of higher isotopes of plutonium, by non-fissile slow neutron capture in ^{239}Pu.

Conversion ratios greater than one can be obtained using fast reactors. Here the core is highly enriched so that the uranium fission chain reaction can be maintained by fast fission neutrons without the necessity of moderators, hence the name 'fast reactor'. With this high enrichment the size of the critical core is quite small and a high proportion of the neutrons may be allowed to escape from the surface to be captured in a larger assembly consisting of depleted uranium surrounding the core. Depleted uranium is uranium from which much of the ^{235}U has been used up by fission in previous reactors and it is here that the breeder reaction

$$^{238}_{92}U + n = {}^{239}_{92}U \xrightarrow{\beta} {}^{239}_{93}Np \xrightarrow{\beta} {}^{239}_{94}Pu$$

takes place. By this means more fissile material can be produced in the blanket than is consumed in the core, a net gain in fissile material. Such reactors are known as 'fast-breeder reactors'.

The spent or depleted uranium breeder blanket may be replaced by the second 'fertile' material thorium where the reaction

$$^{232}_{90}Th + n = {}^{233}_{90}Th \xrightarrow{\beta} {}^{233}_{91}Pa \xrightarrow{\beta} {}^{233}_{92}U$$

produces $^{233}_{92}$U, a uranium isotope which does not occur in nature but, like ^{235}U and ^{239}Pu, is fissile.

A prototype fast-breeder reactor first became operational at Dounreay in Scotland in November 1959. The following details of the first Dounreay fast-breeder, D.F.R., may be of interest.

The small central core of uranium enriched to 46.5% of ^{235}U weighed only 220 kg. The fuel elements were clad in niobium and formed a hexagonal array 21 in. (0.54 m) in diameter and 21 in. long. Some 60 MW of heat was produced within this volume. If one imagines 60,000 one-kilowatt electric fires mounted inside a small dustbin (if space permitted) one has at least an approximate idea of the cooling or heat-transfer problem which such a system presents. The core was cooled not by compressed gas or water, but by a mixture *of molten sodium and potassium which has much better thermal contact and higher conductivity and is therefore more efficient as a coolant. The molten metal is circulated through channels or 'loops' in the core and breeder blanket, using pumps which rely on the propulsive force produced by the interaction of externally applied magnetic fields and a current flowing through the molten metal itself (F = BΛi). These 'magnetic pumps' avoid the difficulties of operating moving parts in a highly corrosive medium. Some fission processes take place in the uranium in the breeder blanket, which contains about 20 tons of depleted uranium clad in stainless steel, and about 12 MW of power is produced in the blanket. The reactor is contained in a double-walled vessel 10 ft 6 in. (3.2 m) in diameter and 21 ft (6.4 m) high and is shown diagrammatically in Fig. 18. The upper floor of the reactor is rotatable and is supported on the biological shield, which is of borated graphite in this case. Steam is raised in the secondary water circuit and 13 Mw(e) of electricity produced.*

Fast breeders of this type, in which the liquid-metal coolant is pumped through the coolant channels or loops in the core and blanket, are known as 'loop-type' reactors. In the D.F.R. reactor a conversion ratio of approximately 1.2 was achieved.

The small D.F.R. reactor operated satisfactorily for almost eighteen years, feeding electricity into the grid and serving as a test facility for the fuel elements and other components of the 250 MW(e) Prototype Fast Reactor (P.F.R.). Having served its purpose it was closed down in March 1977, a year or so after P.F.R. had become operational on the same site.

Although similar in principle P.F.R. differs from D.F.R. in several important respects. The core enrichment is much lower (26%) and is achieved by adding plutonium, and the coolant is molten sodium rather than a sodium potassium mixture. The core is, of course, considerably larger, 1.45 m in diameter and 0.9 m high and is made up of fuel pins 5 mm in diameter clad in stainless steel. The core and breeder blanket are immersed in a large pool of molten sodium coolant weighing 900 tons as shown in Fig. 19.

This type of reactor is known as the 'pool' type as opposed to the 'loop' type of D.F.R. The pool-type coolant system has a very high thermal capacity and this is an important added safety feature to guard against core melt down in the case of failure of the coolant circulation pumps (mechanical in this case) or blockage of the coolant circuits. This will be discussed in more detail in the following chapter.

The operation of P.F.R. was preceded by that of a 250-MW(e) fast-breeder Phénix in France and BN 350, a 'loop'-type F.B.R. in Russia, and all are regarded as prototypes for commercial fast-breeder stations of between 1200 MW(e) and 1500 MW(e) capacity, of which the French 'Super-Phénix' appears to be in the most advanced stage of planning.

Liquid-cooled fast breeders have a breeder ratio approaching 1.2, but, for reasons explained in Chapter 7, a higher breeding ratio is desirable. Initial work is at present in progress in the U.S.S.R. and Western Europe on the design of a gas-cooled fast breeder using helium gas at pressures of up to 100 atmospheres. Paper designs exist for 1200 MW(e) stations with breeder ratios of 1.4 (Vaughan, 1975).

Much more modest breeder ratios, slightly in excess of unity, can be obtained using the thorium ^{233}U cycle in thermal reactors. Work is in progress in the United States of a modified core for P.W.R. reactors which will consist of a central region of fuel pins (the 'seed') containing thorium with a high percentage of ^{233}U, surrounded by a 'blanket' of thorium. In order to avoid neutron loss, and hence reduction of the breeder ratio, the ractor will be controlled not by absorbent rods but by moving the central 'seed' section up and down with respect to the thorium blanket. If successful this core could be fitted into existing P.W.R. reactors at the cost of a 40% reduction in power output. Whilst conversion ratios slightly in excess of unity are possible with ^{233}U-enriched seed this would not be possible with ^{235}U enrichment, so that the reactor depends on an all-thorium cycle.

Candu reactors may also be operated on the thorium cycle to give conversion ratios approaching unity. Conversion ratios of between 0.96 and 1.05 have also been reported from experimental heavy-water reactors using mixtures of thorium and uranium oxide fuel in the form of $5 - 10$-μm diameter spheres suspended in heavy water. These would have the same load-following characteristics as the aqueous solution reactors discussed previously, but would be free from some of the corrosion problems.

The most serious attempt at a thermal breeder reactor to date, however, is the Molten Salt Breeder Reactor which uses a mixture of molten lithium and beryllium fluoride salts for partial moderation, with thorium fluoride salt as the fertile material and either ^{233}U, ^{235}U or ^{239}Pu fluorides as fuel.

These fluorides are all liquid at above 500°C and are pumped through a core region in which graphite provides additional moderation (Fig. 20). The outer regions are under-moderated and it is here that most of the breeding occurs. Conversion ratios of up to 1.07 are possible. A fourteen-year development programme on these reactors is being undertaken at the Oak Ridge laboratory.

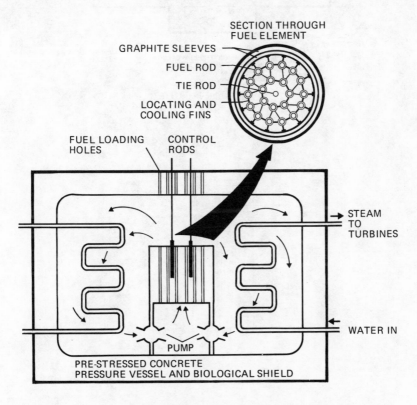

FIG. 13. The advanced gas-cooled reactor (A.G.R.).

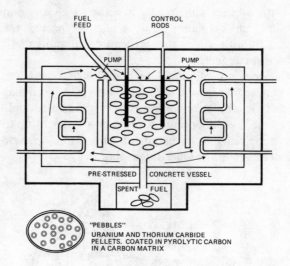

FIG. 14. 'Pebble-bed' H.T.R.

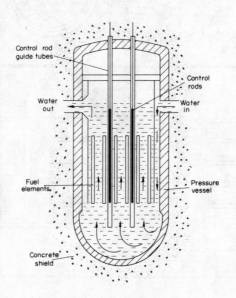

FIG. 15. The P.W.R. reactor 'Yankee'.

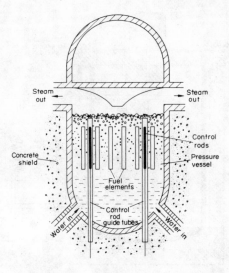

FIG. 16. The boiling water reactor 'Dresden'.

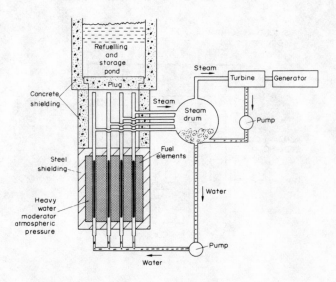

FIG. 17. A S.G.H.W. reactor circuit.

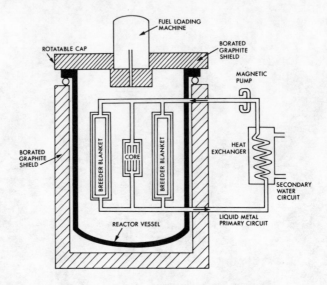

FIG. 18. Loop-type fast breeder (D.F.R. Dounreay).

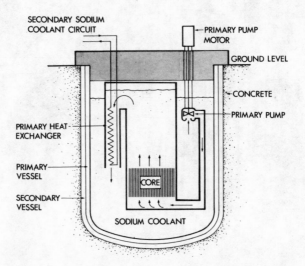

FIG. 19. Pool-type fast breeder (P.F.R. Dounreay).

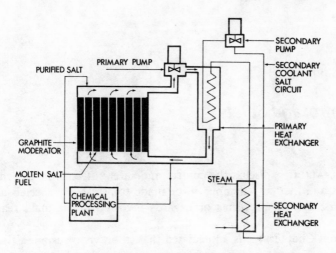

FIG. 20. Molten-salt breeder reactor core and cooling system.

References

CEBONE, R. J. (1973) *Journal of the British Nuclear Energy Society*, 12.4.409.
DEE, J. B., FORTESCUE, D. and LARRIMORE, J. A. (1973) *Journal of the British Nuclear Energy Society*, 12.4.387.
Proceedings of the International Conference on Nuclear Power, Salzburg (1977). Paper IAEA-CN-36/424, L. F. C. Reichle.
ROSENTHAL, BRIGGS and KASTEN (1969) International Atomic Energy Conference Istanbul.
VAUGHAN, R. D. (1975) *Journal of the British Nuclear Energy Society*, 14.2.105.

CHAPTER 6

Control and Safety

WE SHALL first discuss the control and safety of reactors in general terms and then an attempt will be made to assess any special safety factors, control problems or hazards involved in the main reactor types.

From Chapter 4 it is appreciated that in order to sustain a fission chain reaction at least one of the neutrons produced at fission must produce fission in the second 'generation'. That is if the reactor is to run at a constant power output the multiplication factor K must be at least equal to unity and it clearly must be capable of exceeding unity if the reactor power is to be increased, as indeed it must be in order to allow the reactor to be started up from the zero-power (shutdown) condition to its normal working level.

In the initial design of a reactor core it is therefore necessary to ensure that K can be in excess of unity and a certain amount of 'excess reactivity' defined as $\Delta K = (K-1)/K$ is necessary. Since this implies that the number of fission processes can increase from one generation to another and the reactor power can similarly increase continuously, clearly some method of controlling the process to hold ΔK at zero for operation at a steady predetermined power level is required and to reduce ΔK to negative values in order to reduce the power level or even to shut the reactor down. This is in general done by the use of control rods, which are rods of material which capture neutrons very strongly indeed and consequently compete with the fission process for neutron capture. Cadmium and boron are the two most used elements. Since each of these captures low-energy neutrons appreciably more strongly than does $^{235}_{92}U$, a comparatively small amount of cadmium or boron in the reactor core is sufficient to reduce the reactivity appreciably. The number of neutrons in a reactor is proportional to the fission rate and consequently the reactor power, so neutron detectors at prescribed points in the reactor can be used to indicate the power level, as can temperature measurements. In steady operation, if these vary from the readings corresponding to the required operating power, a servo-mechanism withdraws or inserts the control rods to return the power output to the required level. The power level can similarly be altered at will by inserting or retracting the control rods to positions corresponding to the new required level.

Despite all this, the control of a power reactor would be a difficult problem were it not for the extremely fortunate fact that a very small fraction of the neutrons produced as a result of the fission process are emitted, not instantaneously at fission, but some short time later.

In order to appreciate this point let us imagine for a moment that all the neutrons are emitted instantaneously. The time l between one generation of fission processing and the next would be effectively equal to the time taken to slow down (or moderate) the energetic neutron from fission to thermal energies at which it is captured preferentially by $^{235}_{92}U$. This is normally between one-thousandth and one-ten-thousandth of a second and by comparison the time required for neutron capture and the uranium fission process is negligible. If, under these hypothetical conditions, the reactivity were allowed to rise slightly above zero, say $\Delta K = 0.01$, and remain there without correction, even for the longer moderating time, the reactor power would more than double in one-tenth of a second and increase by a factor of over 20,000 in 1 sec $[(dn/dt) = n(\Delta K/l), n = n_0 e^{(\Delta Kt/l)}]$ with consequences which could be disastrous in the absence of any other correcting mechanism. The control of a reactor in which the time between subsequent generations of the fission process was determined only by the short neutron moderating time would therefore require a system of extremely rapid response which would not be easy when one appreciates that control rods are quite heavy and their driving mechanisms correspondingly robust.

Fortunately it is not necessary to overcome this difficult power stabilisation problem because of the 0.65% delayed neutrons referred to in Chapter 4. These are emitted by at least five of the fission product nuclei on average 0.18, 0.49, 2.2, 21.8 and 54.5 sec respectively after the formation of these nuclei.*

Reactors are operated so that if they relied on the neutrons emitted instantaneously on fission, or so-called 'prompt' neutrons, alone, the fission chain reaction would be subcritical, i.e. the reaction would die away. They become critical only if one considers also the 'delayed neutrons'. It is fairly obvious that if the system relies on the delayed neutrons to achieve criticality, the time between one generation of fission processes and the next will depend to some extent on the delay in the emission of these neutrons. The relationship between ΔK and the time required for the reaction to double its power under these circumstances is a complicated one, depending on the moderation time, the average delay of the delayed neutrons and the amount by which the reactor would be subcritical in the absence of delayed neutrons. Table 2 gives typical values. Fortunately under normal operating conditions the power-doubling time is of the order of seconds rather than hundredths of a second and this gives the stabilising mechanisms very adequate time to operate in order to correct even small variations in reactor power.

TABLE 2. *Reactor Power-doubling Time vs. Excess Reactivity*

	Reactor power-doubling time (sec)		
	Neutron moderating time (sec)		
Reactivity (ΔK)	10^{-3}	10^{-4}	10^{-5}
0.001	60	60	60
0.003	10	10	10
0.005	2.5	2.5	2.0
0.006 (near prompt critical)	0.8	0.2	0.14
0.009 (above prompt critical)	0.3	0.04	0.003

It is nevertheless clear that considerations of safety and ease of control require that reactors must not be operated in the 'prompt critical' condition; they must be subcritical if one considers only the prompt fission neutrons chain, and this places an upper limit on the excess

---

*It is probable that there are actually more than five delayed neutron emitters, but reactor control can be explained in terms of these five.

reactivity of say $\Delta K = 0.006$ which would correspond to a multiplication factor due to the prompt neutrons alone of slightly less than unity.

Prompt criticality would be even more serious in a fast reactor of the Dounreay type than in the more usual thermal neutron reactors, since the neutrons do not need to be slowed down between one generation of fissions and the next. The fact that the Dounreay fast reactor could be controlled at all was due to the existence of delayed neutrons, and this is true to only a slightly lesser extent for thermal reactors.

Initially, reactors are fuelled with 'clean' unused uranium, but during their operation the concentration of fission product nuclei builds up and some of these capture neutrons strongly. Neutron capture by fission product nuclei is a competitive process to the fission process itself, but whilst the other competitive processes, capture by $^{238}_{92}U$, capture by structural materials and escape from the surface of the reactor, remain more or less constant with time, the fraction of neutrons captured by fission product nuclei increases as the concentration of these increases, and this must be allowed for in the initial design of the reactor. This point is best illustrated by considering two such fission product nuclei xenon and samarium. Of these, xenon (^{135}Xe) is the more serious, since it captures slow neutrons over 100 times more strongly than does $^{235}_{92}U$. Although $^{135}_{92}Xe$ is formed directly in only 0.3% of fission processes, it is produced indirectly by the successive β-decay of tellurium (^{135}Te) which is produced in 5.6% of the fission processes. ^{135}Xe is itself unstable, decaying first to caesium (^{135}Cs) and then to stable barium (^{135}Ba). The whole process can be expressed as follows:

$$^{135}Te \xrightarrow[\text{2 min}]{\beta} {}^{135}I \xrightarrow[\text{6.7 hr}]{\beta} {}^{135}Xe \xrightarrow[\text{9.2 hr}]{\beta} {}^{135}Cs \xrightarrow[\text{2} \times \text{10}^4 \text{ year}]{\beta} {}^{135}Ba$$

where the times indicate the periods required for half the initial unstable nuclei to transform to the next by emission of β-particles (the so-called 'half-lives' of the β-decay processes).

In fact, therefore, ^{135}Xe is formed from ^{135}Te more quickly than it decays into ^{135}Cs and therefore accumulates in an operating reactor until it reaches a level where its rate of production is equal to its rate of removal by β-decay, plus its removal by neutron capture

$$^{135}Xe + n \rightarrow \, ^{136}Xe$$

to form the heavier isotope ^{136}Xe which does not capture neutrons strongly. When the reactor is shut down, however, the number of neutrons in the reactor core falls to a negligible level and one ^{135}Xe removal process

$$^{135}Xe + n \rightarrow \, ^{136}Xe$$

stops. The ^{135}Xe level then increases, since its half-life for formation (6.7 hr) is shorter than its rate of decay (9.2 hr). This could have the result that if a reactor were not restarted within a few hours of shutdown it would be impossible to restart it until the ^{135}Xe level had fallen due to its own decay to an amount which would allow the overall ΔK to become positive. This could take up to 40 hr, and until then the reactor would be 'poisoned' by the fission product nuclei.

The second main reactor 'poison', samarium (^{149}Sm), is less serious in that it captures slow neutrons only one-tenth as strongly as ^{135}Xe and is formed mainly by the decay of neodymium (^{149}Nd) which is produced in 1.6% of the fission processes. The actual decay process is again a successive β-decay involving promethium as the intermediary nucleus:

$$^{149}Nd \xrightarrow[1.7 \text{ hr}]{\beta} \, ^{149}Pm \xrightarrow[47 \text{ hr}]{\beta} \, ^{149}Sm.$$

As with ^{135}Xe, ^{149}Sm is held at a steady level during reactor operation by the neutron-capture process

$$^{149}Sm + n \rightarrow \, ^{150}Sm$$

and again, like ^{135}Xe, when this removal process ceases on reactor shutdown the level of ^{149}Sm increases, since it is 'fed' by the decay of existing neodymium via promethium. Unlike ^{135}Xe, however, ^{149}Sm is itself stable so that a reactor which became samarium poisoned during shutdown would be impossible to start up until the core or sections of the core were replaced by 'clean' fuel elements. The amount of ^{135}Xe and ^{149}Sm and other reactor poisons which accumulate in reactors either during operation or after shutdown is a function of the reactor type and power, but in the initial design of the reactor using clean fuel, free from fission products, sufficient excess reactivity must be provided to enable the reactor to continue to operate when fission products have reached their equilibrium value at power operation or appreciably above this after reactor shutdown. In the case of Magnox reactors this is of the order of $\Delta K = 2.0\%$ to allow for xenon poisoning and 0.6% to allow for samarium poisoning. In addition, some excess reactivity must be designed into the system so that the reactor will continue to run even after a reasonable fraction (say 40%) of the original ^{235}U has undergone fission and been replaced by a rather smaller amount of ^{239}Pu.

A clean unpoisoned core of the Magnox type of reactor is therefore designed to have an excess reactivity of about 5% in the absence of control rods, but the initial control rod positions are adjusted so that the actual excess reactivity cannot exceed 0.6% in operation to avoid prompt criticality. Enriched reactors, because of their higher power density, produce higher equilibrium levels of fission products and the initial design excess reactivity is correspondingly higher, approaching 10%. Again prompt criticality must be avoided during operation, limiting the actual operational excess reactivity to a maximum value of 0.6%. It is, however, now possible to minimise the effects of fission product poisoning by partially refuelling power reactors during operation and also by 'shuffling' the fuel elements, that is interchanging those which have been exposed to high neutron intensities near the centre of the reactor with those which have been exposed to lower neutron intensities near the edge. In this way fuel consumption and poisoning can be equalised throughout the reactor much as tyre wear on a car may be equalised by occasionally interchanging the road wheels.

The relatively long time constant of a reactor system resulting from limiting the excess reactivity enables reactor control to be achieved by relatively conventional control servomechanism. Failure of the control rods to operate in response to an increase in reactor power is obviously a potential hazard, since, under these conditions, the reactor power would continue to increase if there were no other controlling device.

The first line of defence against such a hazard is a system of ancillary rods of neutron-absorbing material which are normally completely outside the reactor core. These are operated by a separate set of electronic circuits and power-measuring devices which in the case of an increase in power outside the normal limits release these so-called 'shutdown' or safety rods, which then fall onto the reactor and have sufficient neutron absorption to make the reactor subcritical and shut it down. In many reactors the shutdown or safety rods are replaced by small spheres of neutron-absorbing material, since these could enter the core even if it were distorted, and in the case of liquid-moderated reactors neutron-absorbing solutions containing boron salts can also be injected into the moderator by a similar mechanism. It is perhaps unnecessary to add that the electronic circuitry controlling the shutdown mechanism is designed on the 'fail-safe' principle. Circuits are duplicated and often triplicated and a fault on the circuitry itself is designed to shut down the reactor rather than remain undetected. The shutdown rods may also be activated by other potentially dangerous situations, and such hazards are designed either to shut the reactor down immediately or to give visible and audible warning so that corrective action can be taken within a prescribed time before automatic shutdown. The exact procedure depends on the degree of the potential hazard.

There are also inherent factors in most moderated or 'thermal' reactors which would limit the extent of any 'power excursions' in the unlikely event of both control and safety rods failing to operate.

As the reactor temperature increases, both the moderator and the fuel expand and become less dense. The less dense moderator, whether graphite, heavy water or light water, becomes less efficient as it expands and consequently the fraction of neutrons moderated to be preferentially captured by $^{235}_{92}U$ falls. In addition to this, the thermal motion of the $^{238}_{92}U$ effectively broadens the energy interval over which resonant capture by this isotope takes place, the so-called 'Doppler effect' in nuclear reactors. These two factors combine to reduce the excess reactivity which consequently falls to zero, and the reactor power will reach a stable level at an excess temperature which depends on the reactor type and detailed design. The degree of safety provided by this so-called 'negative temperature coefficient' was so high in the case of the prototype homogeneous uranyl sulphate solution reactor described in the previous chapter that it could be used not only as a safety device but also to regulate the reactor power to follow the required load. Neither control nor safety rods are necessary in this particular design.

In other reactor types the negative temperature coefficient is not so high and the temperature rise in the event of malfunction of control and safety rods could reach high values before being limited by the behaviour of the reactor itself. It is therefore relevant to consider the possible consequences of a reactor 'runaway' or 'power excursion'. Fortunately this could not result in a nuclear explosion of the type produced by nuclear weapons, but the heat generated may be such as to melt or even vaporise parts of the reactor fuel and with it accumulated fission product nuclei, most of which are highly radioactive. Of these, the gaseous fission product radioactive iodine (^{131}I) represents the main hazard since it concentrates in the thyroid; radioactive gases xenon and krypton are lesser hazards and the radioactive isotopes of solids strontium (^{90}Sr) and caesium (^{137}Cs) could be hazards, although the escape of solid material is considerably less probable. Melting or vaporising of part of the core would disturb the geometry of the reactor is such a way as to render it no longer critical, and the power level would subsequently fall. The first priority in the event of such an incident is seen as being to limit the spread of radioactive material over the surrounding area and methods of safeguarding against this danger vary from one reactor type to another. In all cases the reactor core is enclosed either in a steel pressure vessel plus a thick concrete biological shield or by a reinforced-concrete container which serves both as a pressure vessel and a biological shield. In the case of most enriched reactor types the whole reactor system, including primary cooling system, biological shield, etc., is enclosed in an additional spherical or cylindrical steel vessel so that the escape of the radioactive material would be confined to the immediate area of the reactor, even if there were an escape of radioactive material from the primary container (Fig. 21).

The only significant release of radioactive material in this country occurred at Windscale in 1958. This was not from a power reactor but from a graphite-moderated air-cooled reactor built specifically for plutonium production. The release occurred when the reactor was in the shutdown state and was due to a phenomenon ('Wigner release') in which neutron irradiation causes the carbon atoms in the graphite moderator to be displaced from their normal position of lowest energy. The graphite consequently stores energy when the reactor is in operation and releases it in subsequent heat cycling of the reactor as the carbon atoms return to their normal positions. In this particular case the stored energy was sufficient to burn some of the graphite and to melt part of the fuel and, since this reactor was air cooled and not equipped with the now standard containment, the fuel melting led to the escape of a certain amount of radioactive material, mainly ^{131}I, to the surrounding countryside. The level of contamination did not represent a direct hazard to life, but was such as to make the milk produced in an area 10 miles by 30 miles unusable due to the accumulation of radioactive iodine (^{131}I) by grazing cattle on contaminated land. Fortunately this fission product nucleus has a half-life of only 8 days and the situation returned to normal within a few weeks.

This previously unsuspected hazard exists only after low-power runs and is not present after reactor operation at normal power since the temperature reached in the moderator in this case is such that displaced carbon atoms have sufficient mobility to return to their normal state during the operation of the reactor. The stored energy in the moderator is therefore much less after a run at normal power than after a run at low power. Although the Windscale incident was the only case of an escape of radioactive material from a reactor in this country, it led to an increase in the severity of the already careful vetting of reactor designs, siting and operating procedures in the U.K. In examining reactor proposals, great imagination and ingenuity are involved in postulating the so-called 'maximum credible accident' where it is assumed that a combination of several unlikely and potentially hazardous faults occur simultaneously and the reactor design and operational procedure must be adequate to limit the escape of radioactive material under this combination of adverse circumstances. In the author's experience the crashing of an aircraft on a reactor building simultaneously with two other only slightly less unlikely fault conditions were combined to constitute a 'maximum credible accident' against which a reactor had to be safe.

In nuclear power, it is by this essentially theoretical approach that hazards are estimated. Fortunately there have been no major reactor accidents which would enable us to compute the likelihood of further such accidents on a purely statistical basis as we do in the case of more familiar hazards such as those involved in road and air transport. From such theoretical studies it is clear that although the reactor power excursion discussed previously is the intuitively obvious hazard, there are other situations which in certain reactor types could present greater potential hazards.

In discussing the heavy-water-moderated natural-uranium Candu reactors in Chapter 4, it was stated that an important added safety feature was the provision of a 'dump tank' for the heavy-water moderator. If all other methods of restricting a power excursion failed, the moderator could rapidly be emptied into the dump tank, and in the absence of the moderator the system becomes subcritical. The superficial resemblance of the pressurised light-water-moderated reactor to the Candu reactors may lead one to believe that the dumping of the moderator could, in this case also, be used as an added safety measure. This, however, is not the case. When a reactor has been in operation for some time an appreciable concentration of fission product nuclei is built up. Six or seven per cent of the total heat produced is due to the radioactive decay of these nuclei and this source of heat is not suddenly removed when the reactor is shut down, either by the insertion of control rods or the accidental loss of the liquid moderator. Because of the enrichment of the light-water-moderated reactors the heat produced by the fission product nuclei per unit volume of fuel is appreciably greater than in the Candu reactors and if the water moderator, which also serves as coolant, were suddenly lost, the reactor fuel elements would melt and form a corrosive molten mass which could possibly penetrate the thick steel pressure vessel. The loss of liquid moderator in this reactor type therefore presents a potential hazard rather than an added safety factor.

The official American view is that the fracture of the steel pressure vessel which would lead to such a release is too unlikely to be regarded as a 'credible accident', and even if it were, the secondary core cooling sprays would be adequate to prevent core melting. This argument has not yet been fully accepted elsewhere, and, in the U.K. in particular, the view is currently taken that the possibility of catastrophic fracture of the heavily irradiated pressure vessel after repeated pressure cycling cannot be completely discounted. The adequacy of the emergency core-cooling system has not as yet been physically proven and more investigation is thought to be necessary before existing American experience can be safely extrapolated to the larger 1200-MW(e) units planned for future programmes.*

Loss of coolant during reactor operation is a potential hazard in most reactor types, but a smaller one in other thermal reactors where the heat produced per unit volume is considerably less. In later Magnox and the A.G.R. reactors the steel pressure vessel has been replaced by prestressed concrete and in all reactors the flow of coolant is carefully monitored.

Fast-breeder reactors present their own safety problems. The heat produced per unit volume of core is even higher than in the P.W.R. so that coolant loss would be even more serious, but since the liquid-metal coolant is at atmospheric pressure this is correspondingly less likely. In the P.F.R. 'pool' type of design the very large thermal capacity of the liquid-sodium pool surrounding the core and blanket would be sufficient to safeguard against core melting due to fission product heating if the coolant circulation system failed. The fast breeder is, however, unique in that if core melting did occur the fuel could conceivably re-form into a configuration with higher reactivity than that of undamaged core. This unlikely eventuality is guarded against by the provision of diverters positioned beneath the core to ensure that a melted core would be split into two subcritical sections rather than one supercritical one.

Safety has always been a paramount consideration in reactor design, and the excellent safety record to date has justified this policy. In the U.K. the Nuclear Installations Inspectorate is responsible for vetting all reactor design and siting proposals, and this body is independent of both the United Kingdom Atomic Energy Authority, the Central Electricity Generating Board and, of course, the commercial firms involved in reactor design and manufacture. Their basic

---

*The recent incident at Harrisburg, U.S.A., illustrates these potential dangers, but it is reassuring that there was no significant release of radioactive material despite several mechanical defects and human errors.

criterion that a reactor design must be such that no single component failure or procedural error can lead to an escape of radioactive material and that the design must be such that a release is only possible if two or three improbable faults occur simultaneously has been quantified. An attempt is being made to estimate the probability of each fault and hence the probability of their simultaneous occurrence. The maximum possible release of ^{131}I which could result from this incident is also estimated and from this the 'Farmer's Acceptable Risk Curve', Fig. 22, is specified (Farmer, 1967). Any reactor design approved by the Inspectorate must be safer than the limits defined by the Farmer Curve. For example, if a particular incident could lead to a release of 100,000 curies* (10^5 curies) the design must be such that the probability of its occurrence is less than once every million reactor years of operation. If the possible release is less, say 100 curies, the probability of its occurrence may be correspondingly greater, once every hundred years in this case.

As indicated previously, one is forced to statistical arguments of this nature due to the happy absence of any significant release of radioactive material from the present power reactor programme. They indicate that the added risk of death or serious injury to a person living within a kilometre of an approved reactor is one in a hundred thousand during a lifespan of seventy years. This is comparable with the risk of being hit by a meteor. Such arguments should be treated with some caution, but they clearly indicate that the risk due to nuclear-power-station malfunction is orders of magnitude less than many of the risks commonly accepted in everyday life, for example, 2% risk of death or serious injury due to road accidents in an average lifespan. Nevertheless, it has been government policy to site nuclear stations in fairly remote areas, with the exception of the two latest A.G.R. stations at Heysham and Hartlepool.

The Nuclear Inspectorate has often been accused of an over-zealous approach within the nuclear industry and their insistence on increasing standards of safety have doubtless contributed to costly delays,

---

*The activity of radioactive materials is expressed in units of the curie, this being equal to 3.7×10^{10} radioactive disintegrations per second, approximately equal to that occurring in 1 g of radium. The effect of radiation in the body will be discussed in Chapter 10.

particularly in the A.G.R. programme, but their obvious independence has done much to facilitate the public acceptance of nuclear-power stations in the United Kingdom.

Nuclear-power stations have not been so readily accepted in other countries. In the United States, for example, the United States Atomic Energy Commission has the dual responsibility of developing the nuclear reactor programme and approving reactor safety and siting. The absence of an obviously independent safety vetting organisation coupled with anxieties expressed by some leading American scientists on the safety of P.W.R. reactors have led to appreciable public opposition to the expansion of the nuclear programme. Despite a few reports of near accidents, the actual operating record to date has been free from any incident causing loss of life (Teller), which is an impressive record in view of the more than 400 reactor years of operation involved. The Rasmussen Report is definite in its assertion that the probable fatality rate associated with the P.W.R. programme is very considerably less than that associated with a comparable coal-fired programme. Nevertheless, it is the P.W.R., and to a lesser extent the B.W.R., amongst the thermal reactors which have attracted the most criticism on safety grounds. The unease which has been typical of the lay reaction in the U.S.A. is being echoed in other countries where, as indicated in Chapter 9, comparatively rapid expansion of the P.W.R.-based nuclear programmes is planned, with perhaps in some cases inadequate public consultation. It is regrettable, however, that the opponents of nuclear power too often cite instances of small accidents and near accidents on early experimental and military reactors which are hardly relevant to the safety of commercial power reactors.

There is still considerable concern regarding the fast-breeder reactor programme although as illustrated by the findings of the 'Flowers Report' this does not devolve mainly about their operational safety but in the management of the plutonium fuel cycle, which will be discussed in Chapter 10.

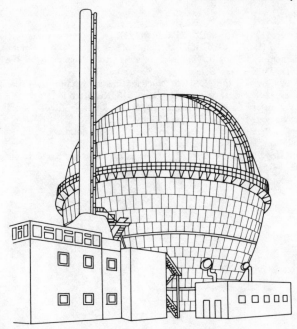

FIG. 21. The outer containment vessel of the A.G.R. prototype reactor at Windscale.

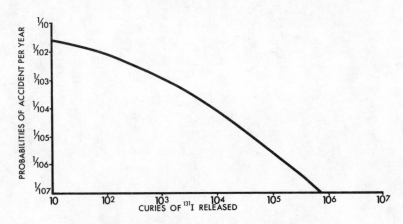

FIG. 22. The 'Farmer' acceptable risk curve.

References

BEATTIE, I. R. (1967) Appendix to Farmer (1967).

FARMER, F. R. (1967) International Atomic Energy Agency. Symposium on the Containment and Siting of Nuclear Power Reactors, Vienna, April 1967.

LEWIS, R. (1973) *The New Scientist*, p. 552, Sept. 1973.

Pollution by Poverty, the Need for Nuclear Energy, Lecture by Professor Edward Teller, U.S. Embassy, London, Oct. 1976. Reported in *Atom*.

The Generic Safety Issues of Pressurised Water Reactors, Report published by the Secretary of State for Energy, 29 July 1977.

The Flowers Report, Sixth Report of the Royal Commission on Environmental Pollution, H.M.S.O., Sept. 1976.

The Rasmussen Report, WASH. 1400, American Nuclear Regulatory Commission, Oct. 1975.

CHAPTER 7

Long-term Economics: The Breeder Reactions and Nuclear Fuel Reserves

IT HAS been shown in Chapter 5 that the use of enriched fuels greatly reduces the constraints placed on the design of nuclear reactors, and this is true whether the fuel is enriched with separated $^{235}_{92}U$ or by $^{239}_{94}Pu$ bred from $^{238}_{92}U$.

Separation of $^{235}_{92}U$ by gaseous diffusion, the centrifugal process or possibly laser excitation is tedious and expensive, but the main argument for using $^{239}_{94}Pu$ rather than $^{235}_{92}U$ for enrichment is not so much this as the necessity to use our fission fuel reserves with reasonable economy. $^{235}_{92}U$ makes up only 0.7% of the total uranium reserves and if reactors use principally $^{235}_{92}U$ as fuel the total economically recoverable reserves of fission fuels are small even in comparison with the fossil fuel reserves. If, on the other hand, an appreciable fraction of the much more abundant $^{238}_{92}U$ can be converted into $^{239}_{94}Pu$ by the breeding process, this can also be used as a fissile fuel, and the total fissile reserves increased by a very large factor.

All uranium-fuelled reactors produce some plutonium, but in order to make significant use of the breeder reaction one must use a reactor with a conversion ratio in excess of unity, a so-called 'breeder reactor' This is best understood by considering a reactor of conversion ratio 0.5, about the lowest found amongst existing power reactor types. In such a reactor for each two nuclei of $^{235}_{92}U$ undergoing fission just one $^{239}_{94}Pu$ nucleus is produced. If we now assume that the plutonium is extracted from the used fuel elements, as it can be by chemical methods, and used in the fuel of a reactor of similar type, we see that, neglecting any plutonium losses in the process, and, in order to illustrate the principle, making the optimistic assumption that all the original fissile material undergoes fission before removal of the fuel element from the reactor, the amount of uranium effectively used in two cycles is:

$$0.7\% \quad + \quad 0.5 \times 0.7\% \quad = 1.05\%.$$
$$\text{initial } {}^{235}_{92}U \quad {}^{239}_{94}Pu \text{ bred in first cycle}$$

If the plutonium bred in the second cycle is extracted and recycled, the utilisation becomes

$$0.7\% + (0.5 \times 0.7\%) + (0.5 \times 0.5 \times 0.7\%) = 1.22\%.$$

The third cycle increases this to 1.31%, a fourth cycle to 1.36%, a fifth to 1.38%, and so on. Clearly the law of diminishing returns is setting in, and if we assume that the process is repeated an infinite number of times, the total percentage of initial uranium eventually undergoing fission is somewhat less than 1.5%. In more general terms the total uranium utilisation is given by

$$\text{utilisation} = \chi (1 + CR + CR^2 + CR^3 \ldots + CR^n)$$

where χ is the percentage of uranium undergoing fission in the first cycle and n the number of cycles. When CR is less than 1 the term in brackets remains small, but it increases to infinity when CR becomes greater than one, increasing the utilisation from a few per cent to 100% in this idealised case. Making due allowance for practical limits to the number of fuel cycles, losses and the build up of higher isotopes of plutonium, Seaborg produced the curve shown in Fig. 23 which shows the same rapid increase in fuel utilisation from a few per cent to over 70% as the conversion ratio increases through unity.

The actual utilisation in non-breeder reactors without recycling, which is the present practice, is also shown. This is approximately 40% of the initial ^{235}U component or approximately 0.3% of the total uranium. It may, therefore, be assumed that the use of fast breeders with fuel recycling would increase the fuel value of our uranium reserves by about a factor of 200 or so, but even this is an underestimate of their actual importance. Although rich reserves of uranium are limited to about 3 million tons (Table 3), uranium is widely distributed on the earth's crust in more dilute form, which is, of course, more expensive to recover. Figure 24 illustrates the pro-

TABLE 3. *Free World Fissile Fuel Reserves*

Uranium. Rich ores recoverable for less than 15 dollars per pound of uranium oxide. In thousands of tons.

Country	Probable reserve	Possible additional reserve
Canada	189	394
U.S.A.	430	655
S. and W. Africa	242	8
France	48	33
Australia	430	104
Nigeria	52	26
Algeria	36	—
Spain	13	11
Argentina	12	20
Gabon	26	6
Other	56	26
	1530	1285

Total 2,815,000 tons.

Thorium. Rich ores recoverable for less than 10 dollars per pound. (Incompletely prospected.) In thousands of tons.

Country	Probable reserve	Possible additional reserve
Canada	90	90
U.S.A.	60	300
South Africa	20	?
Egypt	20	320
Brazil	5	35
India	350	?
	545	745

Total 1,290,000 tons but probably much higher.

bable uranium reserves as a function of refining costs. The costs refer to 1973/74 prices so do not take into account the escalation from 15 dollars per pound to 40 dollars per pound which has taken place subsequently, due in part to general inflation but also to the already threatened shortage of rich reserves.

Even though the economics of present reactor types are not vitally dependent on uranium prices, it seems unlikely that a cost of more than 50 dollars per pound (1973/74 prices) or, say, 150 dollars per pound at today's prices would be acceptable, even if the necessary prospecting to secure these reserves had been carried out. The fuel value of these medium-priced reserves, if used in thermal reactors without recycling, would be about 2 Q or about equal to that of the oil reserves. Used in fast breeders this uranium, with recycling, would have a fuel value of about 400 Q, somewhat more than twice that of the total coal reserve.

More importantly, however, because of the high efficiency of uranium utilisation, fast-breeder economics are even less sensitive to uranium prices and it would be economic to utilise even more dilute reserves, probably including the vast reserves in sea water. The cost of recovering this at a concentration of about 3 parts per billion has been variously estimated at between 200 and 500 dollars per pound, but the total reserve if used in fast-breeder reactors would be over 1,000,000 Q, perhaps 500 times the total coal reserve. Without breeder reactors, therefore, nuclear power must be regarded as a relatively short-term palliative to our energy problem; with breeders it could solve the problem for the foreseeable future. To this must be added the fact that the second possible nuclear fuel, thorium, does not undergo fission directly but, like ^{238}U, can be converted into a fissile isotope ^{233}U by neutron capture and subsequently β-decay, so the utilisation of this fuel is absolutely dependent on the development of breeder reactors.

The thorium reserves have not been fully prospected, but are thought to exceed the uranium reserves by a factor of at least 2. The very rich reserves are shown in Table 3. It will be noted that appreciable reserves exist in India where the need for additional energy sources is acute.

Why, therefore, are we not concentrating on building fast-breeder reactors as rapidly as we possibly can? The problem of public acceptability of the plutonium fuel cycle is discussed in Chapter 10, but, quite apart from this, fast breeders require for their core material uranium enriched to some 20% with plutonium produced from the existing thermal reactor programme, and they can thus only be built in quantity when adequate reserves of plutonium have been accumulated from this programme. This introduces problems of timing and thermal reactor choice which in the opinion of the author have received inadequate attention in the expansion of the existing nuclear-power programme, which has tended to be dominated by the short-term economically competitive position of the various reactor types as discussed in Chapter 8. As a result there appears to be a distinct danger of exhausting the reserves of uranium which are economically viable in thermal reactors before enough plutonium has been accumulated to mount a significant fast-breeder programme.

This can best be assessed by evaluating the uranium consumption and plutonium production from the proposed programmes to the year 2000, by which time we should aim to move to a fast-breeder-dominated system.

Median values for recent forward projections of installed nuclear generating capacity to the year 2000 range between 2.2 million MW(e) and 1.2 million MW(e) (Chapter 9, Fig. 36), and it seems clear that the light-water-moderated reactors will continue to be the dominant type over this period.

It is possible to calculate the uranium-fuel requirement approximately by assuming that 1 ton of natural uranium is required for each MW(e) of installed capacity for the initial reactor charge and 0.3 ton per year MW(e) thereafter for refuelling. The uranium requirement is almost independent of thermal reactor type. It is seen from Fig. 25 (curve a) that the more ambitious reactor programme would exhaust the known rich reserves of uranium before 1985, and the possible rich reserves two or three years later. Thereafter it would be necessary to rely on relatively poor and expensive ores, which are as yet largely unprospected and would possibly be uneconomic to use in thermal reactors.

The less ambitious programme (1.2 million MW(e) by the year 2000) would require a uranium supply shown in curve b of Fig. 25, and even here the known rich reserves would be exhausted well before 1990 and the possible rich reserves by about 1992.

The only way in which the possible rich reserves could be extended to the year 2000 would be by an appreciably more modest programme, the so-called 'conservationist' programme shown as a dotted line on

Fig. 36. This assumes a constant rate of nuclear installation at about 20,000 MW(e) per year globally, and as shown in curve c of Fig. 25 this would just extend the probable rich reserves to the end of the century.

If nuclear power is to have a long-term future we must also examine the plutonium production from these various programmes. Unfortunately, the P.W.R. and B.W.R. reactors are amongst the poorer plutonium producers with a production rate of about 190 kg for one year's operation of a 1000-MW(e) reactor. On the assumption that the most ambitious programme would be based largely on these reactors it would produce only 600 tons of plutonium before the known and possible rich reserves of plutonium became exhausted (Fig. 26, curve a). This would be sufficient to commission less than 200,000 MW(e) of fast-breeder capacity, an insignificant programme compared with the total electricity demand. The 1.2 million MW(e) projection would produce more plutonium before the rich uranium reserves are exhausted, but only slightly so if it were also based on predominantly light-water-moderated reactors. It is only by adopting the more slowly expanding 'conventionist' programme, and basing this on good producers of plutonium such as Candu, Magnox or H.T.R., which produce between 400 and 450 kg of plutonium per year per 1000 MW(e) installed capacity, that adequate reserves of plutonium can be built up before the rich reserves of uranium become exhausted. This programme would produce 2700 tons of plutonium by the time the rich uranium reserves become exhausted, sufficient to mount an 800,000 MW(e) fast-breeder programme* (Fig. 26, curve c).

The total plutonium inventory by the year 2000 would be increased by committing it to fast breeders as soon as it becomes available in significant quantities, say from 1985 onwards. It would, however, be reduced if used in the P.W.R. and B.W.R. programme to ease the strain on uranium separative capacity, as is now happening in Western

---

*It is assumed throughout these calculations that 3.2 kg of plutonium are required per installed MW(e) of fast-breeder capacity, half in the reactor and half being processed at any one time, and that five years' 'cooling' and processing time are required before plutonium produced in a thermal reactor can be used to fuel a fast breeder. For details of the calculations see S. E. Hunt, *J. Inst. Nucl. Eng.*, **19**, no. 2, 1977.

Germany. This will become an increasing temptation elsewhere if the rapid expansion of slightly enriched reactors takes place.

There is thus a clear conflict between the desire to expand nuclear capacity as quickly and cheaply as possible and the necessity, in the longer term, to conserve our uranium reserves, and also build up adequate plutonium reserves to launch a significant fast-breeder programme before these are exhausted. Unfortunately, the short-term considerations appear to be determining the present policy. In contrast to the global situation, the United Kingdom is fortunate in that because of our early start the initial Magnox programme has now produced enough plutonium to fuel two or three fast breeders and it is of interest to note that the fuel value of the stocks of depleted uranium if recycled in fast breeders would be greater than that of the North Sea oil reserves.

Even when a significant fast-breeder programme is mounted, however, there will still be a residual problem of keeping up with the electricity demand. Electricity consumption in the developed countries doubled every twelve years in the recent past. This rate of expansion has been reduced by the oil crisis and clearly cannot continue indefinitely, but a global increase at this rate is clearly necessary if the developing countries are to meet their legitimate aspirations.

The present liquid-metal-cooled fast breeders have a fuel doubling time of about 20 years which would be inadequate to meet this expansion in the electricity demand, particularly if the fast breeders were also required to replace fossil fuel stations to a considerable extent in the twenty-first century. A shorter overall fuel doubling time could be achieved by continuing to supply the fast breeders with plutonium bred from thermal reactors preferably of the near-breeder type using natural uranium such as Candu and Magnox. Further supplementation could be provided if necessary from the isotope separation plant to produce a mixed-reactor system as indicated in Fig. 27.

The alternative to this mixed long-term programme would be the development of a fast breeder with an even higher breeder ratio and a correspondingly shorter fuel doubling time. Work is going on in several countries on the design of a fast breeder cooled by pressurised helium gas. Design studies indicate that this should have a conversion ratio of about 1.4 leading to a fuel doubling time of about ten years.

Fast breeders can, of course, be operated with thorium blankets to produce ^{233}U fuel with much the same breeder ratios as if operated on the plutonium cycle. 'Break-even' or near-unity breeder ratio reactors of the pressurised-water or molten-salt type referred to in Chapter 5 are restricted to the thorium cycle and would be useful in providing additional ^{233}U fuel to a mixed programme including fast breeders, but their breeder ratio is too low to allow them to be the sole basis of an expanding nuclear programme.

In the above discussion it has been assumed throughout that the large reserves of separated ^{235}U and plutonium at present stockpiled for military purposes will not be available for the civil nuclear-power programme. If it were made available two of mankind's major problems could be solved simultaneously.

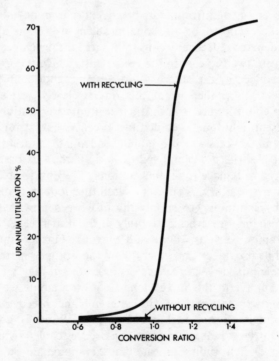

FIG. 23. Total uranium utilisation vs. conversion ratio.

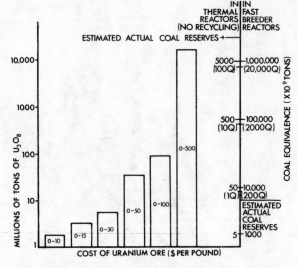

FIG. 24. Uranium reserves as a function of refining costs (1973 – 74 figures).

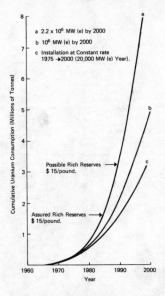

FIG. 25. Estimated uranium consumption to the year 2000.

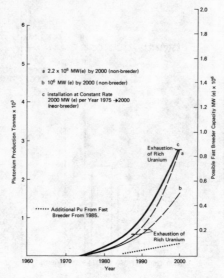

FIG. 26. Estimated plutonium production by the year 2000.

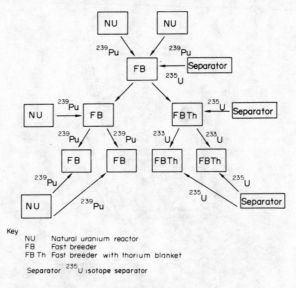

Key

NU	Natural uranium reactor
FB	Fast breeder
FB Th	Fast breeder with thorium blanket

Separator ^{235}U isotope separator

FIG. 27. A possible mixed reactor system for the long term.

References

BARJON, R. (1972) Reactor Technology and Teaching Conference, University of Aston, 1972.

Proceedings of the International Conference on Nuclear Power, Salzburg, May 1977.
Paper IAEA-CN-36/96, E. R. Merz.
Paper IAEA-CN-36/402, P. R. Kaston and F. J. Homan.
Paper IAEA-CN-36/478, R. D. Vaughan and J. Charmanne.

SEABORG, G. T. (1967) *Nuclear Energy*, **16**, Jan./Feb. 1967.

STEINMANN, D. E. (1971) M.Sc. project, University of Aston, 1971.

CHAPTER 8

Short-term Economics: The Cost per Kilowatt Hour

As HAS been pointed out in earlier chapters, the world's need for nuclear power in the long term appears to be inescapable. In the words of the late Dr. H. J. Bhabha, Head of the Indian Atomic Energy Commission, 'No power is as costly as no power', and when the fossil fuel supplies are exhausted a comparison between conventional and nuclear-power costs will become irrelevant. Over-emphasis on ensuring the short-term competitive position of nuclear power may militate against the sensible planning of a long-term nuclear-power programme, but it appears inevitable that in the short term the progress in the installation of nuclear-power stations will depend on the comparative costs in terms of pence per kilowatt hour of electricity from nuclear and non-nuclear stations. Our present civilisation is so dependent on electrical power that the competitive position of a manufacturing nation may well be influenced to an appreciable extent by power-generation costs.

For the early Magnox programme a comparison between generating costs in nuclear and conventional power stations was not a simple matter. The basic difficulty is that in round terms the cost of building a natural-uranium nuclear-power station is about twice that of building a conventional station of comparable output, but against this the running costs of nuclear stations, including fuel and labour costs, are, again in round terms, about one-half those of conventional stations.

In comparing the economies of any two types of station it is, of course, assumed that the capital cost of the station must be borrowed and that interest must be paid on this sum at the commercial rate obtaining at the time, and also that the borrowed capital must be paid back during the operating life of the station. Interest charges and the recovery of the capital costs in a specified number of years therefore becomes a charge against the operation of the station and in the nuclear case this capital charge accounts for some 70% of the total cost of the electricity at present interest rates, compared with rather less than 30% in the case of non-nuclear stations. Thus interest rates and the assumed economic life of the power station are critical factors in making a strictly economic comparison. Clearly if, in both cases, a long operating life is assumed and interest rates are low, the economic argument is biased in favour of nuclear stations, and the opposite assumptions favour non-nuclear stations.

Again, since such a high proportion of the nuclear station costs are due to capital charges, which must be met whether the station is operating or not, an assumption that both stations operate at full power for a large part of their life (i.e. a large load factor) will favour the nuclear stations. For conventional stations the main costs are fuel and labour, both of which are more nearly proportional to the actual output, and the financial penalty for periods of non-operation is less severe.

We have, therefore, three variables—interest rates, assumed economic life and load factor—which are not strictly, or even mainly, dependent on the technical merits of the power station itself, whether it is nuclear or non-nuclear. The interest rates are a function of the general national economic climate of the country in which the power station is built. The assumed economic life of the power station may, of course, be limited by unserviceability due to bad design or construction, but it is more likely to be determined by improvements in the design of future stations which then render it uneconomic by comparison.

The load factor of a power station may also be limited by actual breakdown of the station itself, but it is mainly determined by the fluctuation in the demand for electricity. Clearly, sufficient generating capacity must be installed to meet the peak load and since this is about twice the average load it follows that a large fraction of the generating capacity must be unused for an appreciable fraction of the time, even if perfectly serviceable and capable of operating at maximum output.

We must also add a further uncertainty to our equation by considering the value of the $^{239}_{94}$Pu which is inevitably produced in greater or lesser quantities in all reactors. If the assertions in Chapter 7 are correct, as the author believes them to be, the whole future of nuclear power (and indeed all our future power supplies) may well depend on the accumulation of enriched fuel. In this case, $^{239}_{94}$Pu should be credited at a value equal to the cost of separating $^{235}_{94}$U, or perhaps slightly less because of its more difficult handling properties. In fact it is usually credited for accountancy purposes at appreciably less than this value.

It is clear from the above that by making appropriate assumptions regarding interest rates, economic life of station, duty cycle and the value of plutonium, it is possible to vary the economic comparison of nuclear and non-nuclear power over a wide range. A 1969 comparison of generating costs between the latest Magnox station Wylfa and a modern coal station is shown in Fig. 28. These figures are based on the assumption of a 7.5% interest rate, a twenty-year economic life for the nuclear station and a thirty-year life for the coal station. A 75% load factor is assumed in both cases. On these assumptions the coal station produced electricity at a slightly lower cost than did the nuclear station, but the economic-life assumptions are difficult to justify. There is every indication that Magnox type stations could be reliable producers of electricity for far more than twenty years and since both they and the coal stations will probably become uneconomic, not because they wear out, but when they are superseded by more advanced stations, an assumption of equal economic lives of twenty-five years in each case appears to be more reasonable. Under these conditions the 1969 costs of electricity in terms of pence per kilowatt hour from Wylfa and the best of the coal stations were equal at about 0.27 p/kWh.

In the United Kingdom, therefore, the 'cross over' in the costs producing electricity from the Magnox nuclear stations and the coal-fired stations occurred about 1970 with oil-fired stations slightly cheaper than either of these at that time (Table 4). The decreases in the capital cost of the Magnox stations, which, it will be recalled, is the largest item in the overall costing, fell gradually in real terms from £175 per installed kW(e) for Berkeley in 1962 to about £105 per installed kW(e) for Wylfa in 1969 (Fig. 29). It is important to note that this was not due to any major technological innovation, except perhaps the

TABLE 4. *Nuclear and Conventional Generating Costs in U.K. as at 1969*
(Davis, 1970)

	Capacity [MW(e)]	Year of commissioning	Type	Generating cost (p/kWh)
Calder Hall	35	1956	Magnox	1.0
Berkeley	276	1962	Magnox	0.52
Bradwell	300	1962	Magnox	0.47
Hunterston A	320	1964	Magnox	0.47
Hinkley A	500	1965	Magnox	0.45
Trawsfynydd	500	1965	Magnox	0.40
Dungeness A	550	1965	Magnox	0.32
Sizewell	580	1966	Magnox	0.31
Ferrybridge C	2000	1966	Coal	0.23
Oldbury	600	1967	Magnox	0.31
Tilbury B	1420	1968	Coal	0.30
Wylfa	1180	1969	Magnox	0.29
Pembroke	2000	1970	Oil	0.25
Drax I	1980	1971	Coal	0.27

prestressed concrete pressure vessel, but arose mainly through minor changes based on accumulated experience with some 'economy of scale' resulting from the gradual increase in size of the installations. Due to the need to satisfy the four or five competing consortia companies at that time, by sharing the limited number of orders, the benefits which could have resulted from standardisation and repeat orders were not fully realised. The overall improvement in the economic performance towards a commercially competitive system was, however, impressive as shown in Table 4.

This relatively fine balance between the economics of oil-fired, coal-fired and nuclear-power stations was shattered by the oil crisis in 1973, in favour of the nuclear stations.

Since fuel costs are by far the largest factor in oil-powered stations, generating costs increased about proportionally to that of fuel oil. The cost of generation from existing nuclear stations, because of the relatively low labour content, rose at a rate which was somewhat less than that of the general inflation which followed the oil crisis. Coal stations occupied a central position between these two since the high labour content of the basic fuel meant that coal prices were fully affected by the general inflation and, in fact, rose rather more steeply because of the relatively strong bargaining position of the miners.

Recent trends in the electricity costs in the U.K. are shown in Fig. 30. The balance is now very clearly in favour of the nuclear stations which in 1977/78 had generating costs of a little more than half those of the oil-fired stations and about two-thirds of those of the coal-fired stations. In retrospect it is, of course, a great pity that many more of these then marginally economic Magnox stations were not built in the 1960s.

The attractions of more advanced technology, anticipated lower capital costs and higher thermal efficiency of the slightly enriched reactor types were, however, seductive.

From a costing standpoint slightly enriched reactors are mid-way between fossil-fuelled and natural-uranium stations. Their size for a given power output is considerably less and in principle, therefore, the capital costs which include buildings, etc., should be appreciably lower than for the large Magnox stations. The degree of saving depends on the enriched-reactor type. At one extreme the highly rated P.W.R. reactor is small enough to be mainly factory built so that expensive on site work can be reduced to a minimum, whilst for the A.G.R. the difference is not so marked. If capital costs are lower, however, fuel costs are appreciably higher, since these reactors use fuel which is enriched to about 3% with ^{235}U. This is at present done mainly by the expensive thermal diffusion process, although it is hoped that the gas-centrifuge method may be cheaper and laser excitation, if developed, cheaper still. Present separation costs, however, of about £20 per pound of natural uranium processed approximately double the overall fuel costs compared with natural uranium. Since most slightly enriched reactors have a relatively low conversion ratio the value of the plutonium produced is correspondingly less. The American generating cost figure for P.W.R. stations in 1976 (Crawford), Fig. 28, illustrates this point. Fuel costs are about twice those for the Magnox reactors, although still much less proportionately than those of the coal-fired stations.

The various slightly enriched-reactor types appear to be fairly equally matched in economic terms but it is difficult to obtain valid accurate comparisons between them. In theoretical evaluations varying assumptions on duty cycle, station life and interest rates can have a greater effect on the analysis than do the technical merits of the reactors themselves and in times of high inflation the date of the analysis and the assumptions made in future inflation rates can be the major factor.

A comparative study was made by the C.E.G.B. in 1974 for stations ordered in that year and delivered in 1980. The figures were 0.46 p/kWh for P.W.R., 0.51 p/kWh for S.G.H.W.R., 0.51 p/kWh for H.T.R. and 0.57 p/kWh for A.G.R. Although the absolute figures have little validity now, this at least was a simultaneous study of four systems with common assumptions of a 65% load factor and 10% interest rates. Both assumptions may possibly have favoured the P.W.R. with its assumed lower capital cost and it is possible that the extra costs required to assure public acceptability of the P.W.R. was underestimated. Therefore, there appears to be little to choose in economic terms between the various slightly enriched reactors and, clearly, any system which received sufficient financial backing to be made in quantity would tend to dominate the situation, and this has happened in the case of the P.W.R. reactors. A technical appreciation indicates, however, that they would be expected to be amongst the least promising, with their high fuel costs, poor conversion ratio, limited thermal efficiency and a possible safety problem. The expected lower capital costs may be to some extent illusionary, since construction costs have escalated from $400 ($\sim$£200) per installed kW(e) to $1200 ($\sim$£600) per installed kW(e) between 1972 and 1976 (R. W. Fri). This is a fairly common experience for all reactor types but perhaps a little more extreme than most, since the actual work content in man hours per installed kilowatt has tended to increase rather than fall in the construction of P.W.R.s in recent years despite the increase in average size. This is presumably because of increased attention to safety aspects (Ford Foundation Report). The 1976 P.W.R. generation costs in the U.S.A., however, still show a substantial advantage over average coal-generation costs as shown in Fig. 28, but are actually a little higher than U.K. Magnox costs.

The present very favourable economic position of the Magnox stations is frequently attributed to the fact that they were built before the more extreme escalation in prices occurred in the early and mid-seventies. Since civil engineering costs have risen somewhat more rapidly than the general rate of inflation, the capital cost involved would certainly be appreciably higher now. The higher replacement costs are, however, now reflected in the overall costing of electricity from existing stations and account for some of the apparent rise in Magnox costs in recent years, which is evident in Figs. 28 and 30.

The 1974 C.E.G.B. cost analysis applied to Magnox predicted a capital cost of £366 per installed kW(e) for 1980 completion, giving a generating cost of 0.72 p/kWh. The basic assumption of a 65% load factor was, however, clearly unfavourable to the Magnox case since an average load factor of 81% was achieved over all Magnox stations in 1977.

The five A.G.R. stations in the U.K. have suffered numerous delays which have inevitably led to increased capital costs. Hinkley Point B and Hunterston B are now in operation and although it is too early to give actual costs their initial performance is encouraging (Catchpole and Jenkins).

Costs from the Canadian Candu reactor at Pickering are quoted as 9.8 m$ per kWh or roughly 0.5 p/kWh in 1975 (Foster and Russell) which is approximately equal to average Magnox costs (0.53 p/kWh in that year). Like Magnox, a relatively small fraction of this cost is for fuel since Candu also used natural uranium. Capital costs are less than for Magnox, but heavy-water replacement costs approximately compensate for this.

There is insufficient operating experience on other reactor types to be able to quote realistic operating costs, and arguments will persist regarding their relative economic merits. It seems clear, however, that under most circumstances nuclear power is appreciably cheaper at present than is power from either coal- or oil-fired stations.

Present indications are that the capital costs of fast-breeder reactors would be about 20% higher than for slightly enriched reactors, but fuel costs would be less than half giving an approximately equal overall generating cost (Hamilton and Manne).

As indicated previously, a very large fraction of the total cost of the nuclear stations is in the repayment of capital and interest charges, and a comparatively small fraction in fuel and running costs. Since capital and interest charges must be met whether the station is operating or not there is a strong economic incentive to maintain the duty cycle of the nuclear stations as high as possible. In the case of the fossil stations the main cost is fuel, so that the penalty for non-operation is not so great.

The actual demand for electricity varies by a factor of about 3 over 24 hours, and there is also a considerable seasonal variation. The present practice, therefore, is to meet the steady or base load from the nuclear stations plus the most economic of the fossil stations, and use the less economic of the fossil stations only to meet the additional fluctuations in demand.

This is clearly sensible whilst we have a mixed system of nuclear- and fossil-generating plant, but when the generating capacity becomes predominantly nuclear the fluctuating load will present an obvious problem. Rather than have a nuclear plant lying idle for a large fraction of the time, there will be a strong economic incentive to produce a cheap and efficient energy-storage system. This has already been done at one nuclear site, Trawsfynydd, where any unwanted output is used to pump water from a low-lying reservoir to a high-lying one. This energy is then recovered at times of high electricity demand by a hydro-electric system.*

The energy-storage problem is not peculiar to nuclear plants, and is even more important in attempts to harness intermittent energy such as solar energy, wind and wave power.

---

*A much more ambitious 1650 Mw(e) scheme is under construction at Dinorwic in North Wales. Due to start operation in 1982, it will be the largest pump storage scheme in the world.

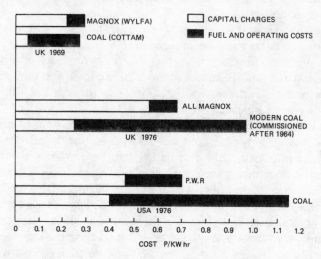

FIG. 28. Coal and nuclear generation costs (not corrected for inflation).

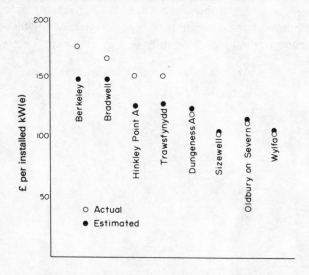

FIG. 29. Estimated and actual capital costs of Magnox stations.

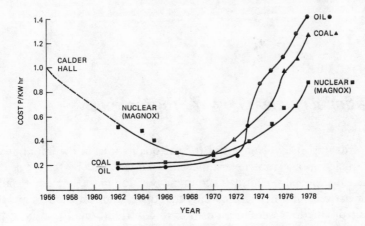

FIG. 30. Trend in electricity generating costs (U.K.) (not corrected for inflation).

References

DAVIS, M. D. (1970) *Nuclear Energy Costs and Developments. Proceedings of Istanbul Symposium.* I.A.C.A. Vienna.

HAMILTON, L. D. and MANNE, A. S. (1978) Health and economic costs and alternative energy sources, *A.E.A. Bulletin,* **20**, No. 4, p. 44, Aug. 1978.

Nuclear Power Issues and Choices. Ford Foundation Report. Bollinger, 1977.

Proceedings of Int. Conference on Nuclear Power, Salzburg, May 1977.

 Paper IAEA-CN-36/397, R. W. Fri.

 Paper IAEA-CN-36/566, W. P. Crawford.

 Paper IAEA-CN-36/53, S. Catchpole and F. P. Jenkins.

 Paper IAEA-CN-36/179, J. S. Foster and S. H. Russell.

CHAPTER 9

National Nuclear Power Programmes

THE degree of involvement of the various nations in the initial phase of nuclear power was determined to a large extent by their wartime nuclear experience and their economic position in the immediate post-war period. This meant essentially that only America, Canada and the United Kingdom were in a position to contribute significantly to the very early stage of nuclear-power development, although Russia and the Western European countries became involved as their economies improved rapidly after the post-war reconstruction period. This was facilitated in Western Europe by the formation of Euratom in 1956, although, as will be seen later in this chapter, this did not prevent individual member countries from pursuing rather different nuclear policies.

The American, Canadian and British programmes were in turn determined by the prevailing economic factors in the three countries in the immediate post-war period. In the U.K. a now admittedly pessimistic forecast of existing fossil fuel supplies indicated that nuclear power might be required to supplement these as a matter of some urgency. The early U.K. programme therefore concentrated on the development of a reactor type which had a high possibility of commercial exploitation in the fairly near future, and this implied the use of natural uranium rather than enriched fuel and this in turn, as indicated in Chapter 4, limited the possible choice of reactor types to two: the heterogeneous graphite-moderated reactor, or the heterogeneous heavy-water-moderated reactor. The former type was selected partly to avoid the high cost of heavy water and partly because appreciable experience in the graphite-moderated reactor had already been accumulated by a number of U.K. scientists and engineers at U.K.A.E.A., Harwell, and at Windscale.

Work on the first natural-uranium graphite-moderated carbon-dioxide-cooled power reactors started at Calder Hall in Cumberland in August 1953 and the first reactor started to feed power into the national grid just over three years later in October 1956. It was followed by three other reactors of very similar design, which reached full power operation between early 1957 and early 1959. The reactors were designed and operated by the staff of the United Kingdom Atomic Energy Authority. Each had a thermal output of 180 MW, but because the maximum temperature of the carbon dioxide gas coolant was limited to 340°C, the efficiency with which this heat could be converted into electricity, the so-called 'thermal efficiency', was only about 19%, giving an electricity output of some 35 MW(e) per reactor.* The Calder Hall reactors were followed by four more of the same type, still designed by the U.K.A.E.A., at Chapelcross in Dumfriesshire. Minor changes increased the heat output to 225 MW per reactor and a slightly higher gas output temperature increased the thermal efficiency to nearly 20%, giving an electricity output of 41MW(e) per reactor.* The reactors came on to full power at various dates between early 1959 and early 1960. The success of the Calder Hall reactors was such that they were considered suitable to form the prototype of the first series of large nuclear-power reactors, the 'Magnox' reactors, and work on the first two commercial nuclear generating stations at Berkeley in Gloucestershire and Bradwell in Essex started in early 1957. The stations, each of which has two reactors, were designed and built by two of the newly formed 'nuclear consortia' for the Central Electricity Generating Board. The Berkeley station was the responsibility of the 'A.E.I.—John Thompson Nuclear Energy Company Ltd.', and that at Bradwell was built by 'The Nuclear Power Plant Company'. Although the reactors were of the same general type as those at Calder Hall they were appreciably larger, the heat outputs being about 550 MW per reactor and by raising the maximum gas temperatures to about 390°C thermal efficiencies of between 25% and 28% were achieved to give electricity outputs of about 150 MW(e) per reactor. Unlike the prototype, the Berkeley and Bradwell reactors were designed so that the reactor fuel could be

---

*These reactors were subsequently up-rated to 268 MW(t) and 50 MW(e) per reactor.

changed in sections without shutting the reactor down. The first reactors of each station commenced operation in early 1961 and the second a year later.

Another station consisting of two 160 MW(e) Magnox reactors was completed at Hunterston in Ayrshire in 1964 by the 'G.E.C. Simon Carver Atomic Energy Group' and larger stations each consisting of two 250 MW(e) reactors were completed at Hinkley Point, Somerset, and Trawsfynydd, Merionethshire, in 1963 and 1964 respectively. Slightly larger, two-reactor stations were completed at Dungeness in Kent (total output 520 MW(e)) in 1965, at Sizewell in Suffolk (580 MW(e)) in 1966, and Oldbury in Gloucestershire (600 MW(e)) in 1967. The last of the Magnox series, Wylfa in Anglesey (1180 Mw(e)), was completed in 1969. All these reactors were commissioned by the Central Electricity Generating Board and built by one or other of the nuclear consortia companies, the companies themselves undergoing appreciable regrouping during this period. Although the reactors were of the same basic type, there was a gradual, almost evolutionary, development throughout the series. Relatively minor design improvements reduced construction costs and increased the carbon dioxide output temperature and with it the thermal efficiency, resulting in a gradual improvement in the economy of the stations. Perhaps the major development was the introduction of prestressed concrete containment in Oldbury and Wylfa to replace the steel pressure vessel of earlier reactors.

In passing, it should be noted that reactor installations are all situated near plentiful water supplies, because the thermal efficiency of about 30% implies that 70% of all the heat produced must be discharged ultimately to surrounding rivers or sea. In this respect the Magnox reactors are slightly worse than the fossil fuel stations, the best of which achieve a thermal efficiency of about 40%.

With the completion of Wylfa the Magnox (or Mark I stations as they are also known) reached a total electricity-generating capacity of 5000 MW(e), representing about 12% of the total generating capacity of this country.

The next generation of power reactors aimed essentially to attain a lower capital cost and higher thermal efficiency by the use of slightly enriched fuels. The Advanced Gas Cooled Reactors (A.G.R.s), also known as the Mark II gas-cooled reactors, effectively started with the prototype reactor at Windscale in Cumberland. Constructional work started on this reactor in late 1958 slightly after the successful operation of the Magnox prototype at Calder Hall and the 33-MW(e) reactor was completed in 1963. The general characteristics of this reactor type are described in Chapter 5. Five stations of this type were ordered at Hunterston, Hartlepool, Hinkley Point, Dungeness and Heysham. Each of these has a designed electricity output of 1250 MW(e) and two stations are now in operation (Fig. 31).

The construction of the A.G.R. stations suffered from long and costly delays in which the nuclear aspects of the A.G.R. design played only a minor role. Comparable delays and cost escalation were also experienced on coal-fired stations under construction during this period.

The delays in the A.G.R. programme, however, were one of the factors which produced considerable uncertainty in planning the next phase of the United Kingdom programme. Possible choices included the S.G.H.W.R. and H.T.R. reactors which had been developed at Winfrith at least to the 100-MW and 20-MW level respectively. Consideration was also given to building a series of P.W.R. reactors using essentially American expertise and of collaborating with Canada, either by adapting the Candu system or by using their experience to develop the similar S.G.H.W.R. system to commercial power levels. A further series of the then unpopular A.G.R. stations represented the sixth possibility.

This rather bewildering range of options was further complicated by what turned out to be over-provision of conventional stations in view of the competition from North Sea gas in the domestic market and the overall recession in the early and mid-1970s. Against this the once thriving nuclear-power industry had been starved of new orders for over eight years.

Opinions on reactor choice were divided: the C.E.G.B. expressed a preference for the P.W.R. system and the Southern Scotland Electricity Board (S.S.E.B.) for S.G.H.W.R.

Despite the indecisive outcome of the previous Vinter Committee the problem was again referred to a select Parliamentary Committee in 1974 and, influenced to some extent by doubts on the P.W.R.

safety expressed, amongst others, by Sir Allen Cottrell, their Chief Scientific Advisor, the Committee recommended that the S.G.H.W.R. should be adopted. Plans for building two such stations were put forward, but this decision was reversed two years later due in part to difficulties in extending the S.G.H.W.R. reactor design to a commercial size, but mainly to the successful commissioning of the A.G.R. stations at Hinkley Point and Hunterston in 1976. This led to the decision to build two further A.G.R. stations, but to keep an option open on P.W.R. reactors should their safety assessment by the Nuclear Installations Inspectorate prove to be favourable. Development work on the H.T.R. and S.G.H.W.R. was consequently discontinued or severely curtailed.

During this period considerable reorganisation took place within the nuclear-power industry, and the consortia companies, which at one time numbered five, were reduced to one, the National Nuclear Corporation in which the Government have a 35% holding.

Britain's performance after the Magnox series of reactors was frankly disappointing due in part to indecision and organisational problems and in part to the unexpected competition from North Sea gas, but a good start was made on what is now generally regarded as the essential component of the long-term programme, the fast breeders.

In 1960 a small 13-MW(e) experimental fast breeder (D.F.R.) was built at Dounreay in Scotland and operated very satisfactorily until 1977. In addition to feeding power into the grid it served as a test facility for the fuel elements and other components of the 250-MW(e) Prototype Fast Reactor (P.F.R) which succeeded it at Dounreay and became operational in 1976. It is anticipated that plans for a commercial 1300-MW(e) Fast Reactor (C.D.F.R 1) will shortly be submitted for planning permission at the Dounreay site, where a reprocessing plant for fast reactor fuel is nearing completion. The long-term U.K. nuclear programme, therefore, appears to be reasonably well assured, particularly as adequate stocks of plutonium will be available due to our early start with Magnox reactors.

The total installed nuclear capacity in the U.K. for the year 2000 is now planned at 40,000 MW(e) which will then meet about 25% of our total electricity demand, as illustrated in Fig. 32. This represents a considerable reduction in earlier plans up to 1990, but thereafter a rapid increase in nuclear capacity is anticipated to meet the shortfall as North Sea gas becomes exhausted and to replace conventional plant, many of which will become obsolete about that time.

In contrast with the immediate post-war situation in this country, it seemed at the outset that there was no immediate shortage of fossil fuel in the U.S.A. and if nuclear power was to play a significant part in the overall power programme it must be commercially competitive in terms of costs per kilowatt hour. This led to an experimental approach with a wide variety of reactor types, taking advantage of the greater design flexibility offered by the use of enriched fuel produced in the extensive isotope separation plants which had been part of the military programme. Separated $^{235}_{92}U$ could be regarded as almost a 'war surplus' material which was leased at attractively low rental by the U.S. Government for reactor-development work. The American effort was not solely devoted to the investigation of reactor types suitable for large power-producing reactors, but also included the development of smaller reactor systems for ship, submarine and rocket propulsion. Their wider experimental approach to the investigation of possible power reactor types has included the natural heterogeneous graphite-moderated, the slightly enriched heterogeneous graphite-moderated, the slightly enriched heterogeneous light-water-moderated reactors of the pressurised and boiling-water types, more highly enriched small P.W.R. reactors, reactors using organic liquids, heavy-water-moderated reactors, fast breeders of the Dounreay type, high-temperature helium-cooled reactors similar to the H.T.R. at Winfrith, homogeneous reactors of the aqueous solution type and molten salt reactors.

With the exception of a relatively small 40-MW(e) H.T.R. reactor at Peach Bottom, Philadelphia, which was commissioned in 1967, and a larger 330-MW(e) of this type which became operational in 1977 at Fort St. Vrain, Colorado, development of reactors to power reactor stage has been confined to the slightly enriched light-water-moderated types. These are to some extent modifications of reactors initially developed for submarine propulsion as part of the military programme. The 200-MW(e) B.W.R. prototype 'Dresden' was completed in 1960 and the 174-MW(e) P.W.R. prototype 'Yankee' a year later.

Shippingport, also a P.W.R., was operational four years earlier, but this was rather different in concept, having a highly enriched central core and a breeder blanket.

The P.W.R.s are designed and manufactured by Westinghouse, Combustion Engineering or Babcox and Wilcox and the B.W.R.s by General Electric. Compared with the British programme, few nuclear-power stations were commissioned in the early 1960s. Apart from those already mentioned, only a 265-MW(e) P.W.R. at Indian Point, New York State (1963), was commissioned before 1965. A 790-MW(e) P.W.R. station also became operable in Washington in 1966. Three reactors with a total generating capacity of about 1000 MW(e) came into operation in 1967 and none in 1968, but the large expansion in the nuclear programme started in 1965 and 1966 resulted in a greatly increased generating capacity commissioned in 1969 and 1970.

The total installed nuclear capacity in the United States equalled that of the United Kingdom at 5000 MW(e) in 1970, but subsequent expansion has been rapid, to 52,000 MW(e) in 1977 with estimates of between 130,000 and 160,000 MW(e) by 1985. Estimations range between 370,000 and 810,000 MW(e) by the turn of the century. Up to 1990 at least this rapid expansion will continue to be based mainly on P.W.R. and B.W.R. reactors, built by Westinghouse and General Electric respectively.

This programme is recognised to be costly in uranium and the domestic programme will exhaust the known American reserves by 1990. It is being accompanied by extensive uranium exploration and the construction of additional isotope-separation plant using the gas-centrifuge system.

The Carter Administration has, for the time being at least, placed a moratorium on fuel recycling and with it the 'plutonium economy'. Spent-fuel elements will be stored in their unprocessed state and considerable effort is being devoted to the location of safe storage areas.

The American contribution to the fast-breeder programme had been considerable, starting with the experimental breeder-reactor cores at Los Alamos as early as 1946, and progressing to E.B.R. 2, the 37.5-MW(th) core at Idaho Falls in 1964, with a parallel effort on the Enrico Fermo 200-MW(th) reactor from 1955 to 1973. With the present rejection of the plutonium economy this once considerable effort has been curtailed and the 975-MW(e) fast-breeder reactor planned for the Clinch Point site in Tennessee has been cancelled.

It is, however, recognised that in the slightly longer term plans must be made to improve the uranium utilisation, and to this end additional effort is being devoted to the fused salt 'break even' thermal breeder using the thorium cycle, and the development of modified breeder cores for the P.W.R. reactors. The effectiveness of these measures is discussed in Chapter 7.

The third immediate post-war contributor to the development of nuclear power, Canada, also had appreciable fossil reserves and cheap hydro-electric power. This cheap hydro-electric power made it possible for them to produce heavy water reasonably cheaply and hence to develop the heavy-water-moderated and cooled reactor 'Candu', which has been the basis of their programme since commissioning of the small 22-MW(e) Nuclear Power Demonstration Reactor (N.P.D.) in Ontario in 1962. This was followed by the first (208 MW(e)) commercial Candu at Douglas Point in 1967 and then by four 514-MW(e) reactor stations, Pickering A, in Ontario, commissioned between 1971 and 1973. A 250-MW(e) reactor, Gentilly 1, in Quebec ir 1971 concluded the domestic installations for a time but a 125-MW(e) station was exported to Pakistan in 1971 and two 200-MW(e) reactors to India in 1972 where they have become the basis of the Indian programme. The first of a further four reactor stations, Bruce A, with total capacity 3000 MW(e), in Ontario started operation in 1976 and the station should be at full power by 1979, to be followed by a 600-MW(e) reactor, Gentilly 2, in Quebec. Plans to 1990 include a 600-MW(e) station at New Brunswick and then large four-reactor stations, Pickering B, Bruce B and Darlington in Ontario, plus stations for export to Argentina and Korea. The firmly planned domestic programme to 1990 totals 15,000 MW(e) plus 1750 MW(e) for export. The success of the Candu programme appears to illustrate the wisdom of confining the national effort to one reactor type.

As previously indicated, the immediate post-war conditions on the continent of Europe prevented most of its countries from making an immediate contribution to the nuclear-energy programme, but the delay was surprisingly short. By 1954 a 38-MW graphite-moderated air-cooled natural-uranium reactor G_1 was under construction at Marcoule in France. It was built primarily for the production of plutonium and was operational early in 1956. It was followed by two others, G_2 and G_3, of the same general type, but larger. Each had a thermal output of 260 MW and carbon dioxide cooling replaced the air cooling used in G_1. These reactors produced 40 MW(e) of electricity each as well as plutonium. All three reactors were commissioned by the French Atomic Energy Commission, and G_2 and G_3 were operated jointly by the Commission and 'Electricité de France'. They had horizontal rather than vertical fuel rods and the G_2 and G_3 reactors had octagonal cores. They resembled Calder Hall in that they were of the same general type and comparable in size, but the detailed design and overall appearance were very different. To accommodate the horizontal fuel rods the graphite moderator was in the form of a horizontal octagonal prism and even at this early stage the steel pressure vessel was replaced by a horizontal prestressed concrete cylinder which served as both pressure vessel and biological shield. The control and safety rods were vertical.

The first 'commercial' French nuclear-power station, 'Centrale de Chinon EDF 1', reached criticality in 1962, just a year after its British counterparts at Berkeley and Bradwell. Its general appearance was, in fact, much closer to the British reactors than to the French G_2 and G_3 'prototypes' in that the core was a vertical cylinder and the apparently retrograde step of using a steel rather than a prestressed concrete pressure vessel was taken. This reactor was smaller than its British counterpart, having an output of 300 MW(t) and 68 MW(e), but it was followed by a larger reactor Chinon EDF 2 of output 791 MW(t) and 198 MW(e) (1965). This was again of the general 'Magnox' or 'Graphite-gaz' type and included the refinement of continuous refuelling during operation. The pressure vessel was spherical in shape and consisted of welded steel plates. The Chinon reactors were designed by the French Atomic Energy Commission and 'Electricité de France' and owned and operated by the latter. Although private industrial firms were very much involved in the construction they did not take overall responsibility as did the British nuclear consortia firms. Nevertheless, the early phases of the nuclear programme in France and the U.K. followed closely parallel but largely independent courses. Their preference for the graphite-moderated carbon-dioxide-cooled reactor was shown by their commissioning of another power station of this type, EDF 3, at Chinon in 1966. This station has a total generating capacity of 480 MW(e) and was followed by similar stations at St. Laurent, St. Laurent I (1969) and II (1970) with generating capacities of 480 MW(e) and 515 MW(e) respectively. The series was terminated by the construction of a slightly larger station at Bugey with an output of 540 MW(e) in 1971. As in the U.K. the end of the natural-uranium graphite-moderated series produced a problem.

A relatively small 73-MW(e) enriched heavy-water-moderated gas-cooled reactor had been built at Brenniles in 1966, but rather surprisingly in view of the political situation about this time, Electricité de France then invited tenders for P.W.R. and B.W.R. reactors. A 900-MW(e) P.W.R. was ordered to become operational at Fessenheim in 1976. Their flirtation with the American light-water-moderated systems was appreciably accelerated by the oil crisis of 1973, which was much more serious for France because of its depleted fossil fuel reserves. In 1974, therefore, the previous forward plan for nuclear generating capacity by 1985 was increased from 30,000 MW(e) to 45,000 MW(e) and, if achieved, this will represent 72% of the total electricity-generating capacity at that date. The programme will be made up mainly of 900 MW(e) P.W.R. stations constructed by Framatome under licence from Westinghouse. The reactors will be grouped in a number of four reactor stations, and their size may be increased to 1300 MW(e) each towards the end of the series. A smaller number of General Electric B.W.R. reactors will complete the programme.

This massive increase in the light-water-moderated reactor programme has met with considerable public opposition which could be well be one cause of delay. The overall uranium supply and the provision of adequate isotope-separation capacity are other obvious difficulties. The French are, however, very much aware of the necessity of longer-term planning and have declared their intention of processing their spent fuel to provide plutonium for their fast-breeder programme, which is well advanced. The 250-MW(e) Phénix reactor became operational in 1975, almost a year before our own P.F.R., and a decision has been reached to build a commercial 1300-MW(e) fast-breeder Super Phénix, in which Western Germany and Italy will have minor interests.

In Western Germany the initial phase of reactor development showed a marked American influence in the use of enriched fuel which made possible a rather broadly based experimental approach. Their first reactor, a small 15-MW(e) B.W.R. at Karl in 1961, was built as a joint project between General Electric of America and Allgemeine Elektrizitäts-Gesellschaft (A.E.G.). The experimental approach was maintained in the building of a heavy-water research reactor at Karlsruhe (critical in 1966), a 15-MW(e) pebble-bed H.T.R. at Julich in 1967 and a prototype fast breeder at Karlsruhe in 1972.

The first phase of the commercial reactor programme, however, was dominated by light-water-moderated reactors beginning with a 250-MW(e) B.W.R. at Gundremmingen in 1966 and followed by a 280-MW(e) P.W.R. at Obrigheim in 1968, a 240-MW(e) B.W.R. at Lingen in 1968, a 630-MW(e) P.W.R. at Stade in 1971 and a 640-MW(e) B.W.R. at Wurgassen in 1972, and with the introduction of a larger B.W.R. station at Brunsbuttel and a large P.W.R. at Biblis this brought the total nuclear generating capacity in Western Germany to 3500 MW(e) by 1975, with B.W.R. stations built by A.E.G. and P.W.R.s built by Siemens under licence from Westinghouse playing approximately equal roles.

Plans were made in 1976 to increase this capacity to 35,000 MW(e) by 1985, using mainly P.W.R. reactors, but these have met with a more violent adverse public reaction than in France, which is almost certain to curtail or delay their programme considerably.

Like the French, the Germans intend to continue to process their spent-fuel elements and are already using some of the plutonium extracted to enrich fuel elements for their P.W.R. programme, thus reducing the demand on uranium isotope operative capacity. They are, however, fully involved with the Netherlands and the United Kingdom in the collaborative gas-centrifuge-separation programme.

Alongside their commercial light-water-reactor programme the West Germans are maintaining their interest in the pebble-bed version of the H.T.R. reactor and are seeking further international co-operation in the development of this for electricity generation and as a direct heat source for industrial processes.

Their long-term programme will certainly include fast breeders, and following the small 1972 Karlsruhe reactor a loop-type 300-MW(e) demonstration plant (S.N.R. 1) is planned for completion at Kalkar in 1981. This is a collaborative project with Belgium and the Netherlands. It is intended that the Western German interest in Super Phénix will be followed by a fast breeder of similar size (S.N.R. 2) in Germany.

Italy's first nuclear reactor, 'Latina', was a 200-MW(e) Magnox built by The Nuclear Power Company Ltd. in collaboration with the Italian firm A.G.I.P. Nucleare. It reached full power in 1962. This was followed by B.W.R.s and P.W.R.s of comparable size at Garigliano and Varcelli in 1963 and 1964 respectively. An 850-MW(e) B.W.R. at Mezzanone di Caorso became operational in 1977 and a further four 1000 MW(e) stations are on order. It is planned to increase the nuclear generating capacity to about 10,000 MW(e) by 1985. This will be made up of mainly B.W.R. and P.W.R. reactors.

Smaller Western European countries have in general started with modest nuclear programmes based mainly on the light-water-moderated reactors. These include two 880-MW(e) P.W.R. stations in Belgium at Doel (Antwerp) and Tilange (Huy), which came into operation in 1974 and 1975. Switzerland has two 350-MW(e) P.W.R. reactors at Beznau (Doettingen) which came into operation in 1969 and 1972 and a 300-MW(e) B.W.R. near Berne which was completed in 1971. A 400-MW(e) P.W.R. became operational near Flushing in the Netherlands in 1975.

Sweden has a relatively large nuclear-reactor programme, starting with a 140-MW(e) slightly enriched S.G.H.W.R. type reactor at Marviken in 1969, followed by B.W.R. and P.W.R. stations which now have a generating capacity of 3200 MW(e). This programme is planned to grow to 10,000 MW(e) by 1985. Spain also plans a rapid expansion of its existing programme to about 14,000 MW(e) by 1985. The present Spanish generating capacity consists of a 150-MW(e) P.W.R. which has been in operation near Madrid since 1968, a 440-MW(e) B.W.R. at Burgos and a 480-MW(e) graphite-gaz (Magnox type) built in collaboration with France at Tarragona.

The nuclear programmes have been discussed so far in terms of the installation of commercial power stations, and with the emergence of the slightly enriched reactors as the dominant type in the middle term, increased isotope-separation capacity becomes a vital part of the overall programme. Diffusion-type separators at Oak Ridge and later Capenhurst are extremely costly both to construct and to operate, and the development of a gas-centrifuge isotope-separation system has been a collaborative venture between Western Germany, the Netherlands and the United Kingdom. The Capenhurst plant has been in operation since September 1977, as has a similar plant at Almelo and each represents the first phase of a much larger development.

Fuel reprocessing is another major requirement, and following the Windscale Enquiry British Nuclear Fuels now have authority to build a large spent-oxide fuel-reprocessing plant at Windscale. This will serve the U.K. and several overseas countries. A similar plant with international commitments is expected to be operational in Karlsruhe in Western Germany in the late eighties.

The Russian nuclear programme has not involved direct collaboration with the Western powers, and consequently has resulted, at least in the early years, in rather different reactor concepts. Endowed with very considerable fossil fuel resources they could afford to make an experimental approach, yet they can claim to have produced their first nuclear-generated electricity. This was from a small 30-MW(th), 5-MW(e) graphite-moderated reactor using 5% enriched fuel and cooled by pressurised water. It could be operated in either the boiling or non-boiling condition and first fed power into the grid at Obninsk in late 1954, almost two years before Calder Hall. Their first large scale station, however, did not appear until 1964. This consisted of four 100-MW(e) graphite-moderated water-cooled reactors at Beloyarsk in the Urals. The fuel in this case was enriched to only 1.5%. Two more conventional boiling-water reactors, each of 250-MW(e) capacity, were completed at Veronezh in the same year with another similar 50-MW(e) unit at Ulyanovsk. The Russian programme until the late eighties will be based on a mixture of P.W.R.-type reactors and the graphite-moderated water-cooled B.W.R. reactors, in approximately equal numbers. The present installed capacity is 4600 MW(e), planned to increase to 14,400 MW(e) by 1985.

In parallel with their relatively modest commercial thermal-reactor programme the Russians have maintained a sustained interest in the breeder-reactor concept. A small plutonium-enriched 5-MW(th) fast reactor reached criticality at Obninsk in 1958 and was followed by a 60-MW(e) liquid-sodium-cooled experimental fast breeder at Melekess in 1969, which appears to have operated satisfactorily since then, apart from two small sodium leaks. The larger B.N. 350, a loop-type fast breeder, was commissioned in 1973 at Kazakh and as well as producing electricity, is used in a pilot desalination plant for water from the Caspian Sea. It is designed to produce 150 MW(e) of electricity and 120,000 m³/day of fresh water, but has so far achieved only about 30% of this output.

The latest Russian breeder, B.N. 600, a 600-MW(e) pool-type fast breeder, is approaching criticality and is claimed to be the first 'commercial-sized' fast breeder. It will be the basis of the design of a proposed 1500-MW(e) station.

All these are sodium-cooled fast breeders capable of being used with either ^{235}U or ^{239}Pu enriched cores. The Russians are also expending considerable effort on the gas-cooled fast-breeder concept, which with its higher breeder ratio will allow a more rapid expansion of the programme as discussed in Chapter 7.

Japan was understandably a late starter in the nuclear race. A demonstration 12-MW(e) B.W.R. supplied by General Electric of America came into operation in 1963 at Tokai-Mura and a 160-MW(e) Magnox supplied by the British General Electric Company was commissioned at the same location in 1965. The subsequent programme, based largely on light-water-moderated reactors manufactured under licence by Japanese industry, expanded rapidly to 13,000 MW(e) in 1978 and is planned to reach 50,000 MW(e) by 1985, and three times this by the turn of the century. Although light-water-moderated reactors will dominate in the intermediate time scale the plan is to move to fast breeders in the longer term, and the construction of a 300-MW(e) loop-type breeder, Monju, is due to start next year with planned operation for 1986.

Amongst the developing countries the Indian and Pakistan preference for Candu reactors has already been referred to. In India two 198-MW(e) B.W.R.s were commissioned in 1969 and are operating at Taranpur, but they were followed by a 207-MW(e) Candu at Rajasthan in 1973 built with Canadian co-operation and this will shortly be followed up by a second Indian-built Candu at the same site. A further four 220-MW(e) Candu stations are under construction for planned operation by 1982. Pakistan has one 125-MW(e) Candu operating at Kanupp and a 600-MW(e) Candu planned for Chasnupp. Argentina also has one 319-MW(e) Candu, Atacha 1, which has been in operation since 1974 and two 600 MW(e) Candus under construction. Elsewhere initial programmes are based on P.W.R. reactors usually supplied under assistance agreements by the United States. P.W.R. stations are under construction in Brazil, Egypt, Iran, Korea and the Philippines with two B.W.R. stations in Mexico. The Communist countries outside the U.S.S.R. have also, in general, opted for P.W.R. reactors built with Russian assistance, rather than the alternative graphite-moderated water-cooled reactors also available from Russia. Two 420-MW(e) P.W.R. reactors are in operation in Bulgaria and P.W.R.s are under construction or planned in Cuba, Czechoslovakia, Hungary, Poland and Romania.

Actual and planned growths of some of the national nuclear programmes are shown in Fig. 33.

The pattern for choice of reactor type in the medium term is now firmly established. The early development to commercial sizes of the natural-uranium graphite-moderated reactors in the U.K. and France led to significant installed capacities in both these countries, giving them an early lead by 1970, Fig. 34. This reactor type had much to commend it, but its short-term competitive position before the oil crisis was prejudiced by relatively high capital costs, the effect of which was accentuated by the high interest rates ruling generally in the 1960s. Although they did become commercially competitive with fossil fuels towards the end of this period few were exported.

After an experimental approach over a broad field the Americans opted for the light-water-moderated reactors in which the P.W.R. type predominated and their domestic reactor-construction programme became significant in 1965, leading to a rapid increase in installed capacity from 1970 onwards. This increase in domestic installations was accompanied by an intensive drive to export this type of reactor to Western European countries and to the developing countries, often with assistance from the American government under international aid agreements, and the light-water-moderated reactors are now the dominant reactor type globally (Fig. 35). These reactors are well suited to the export market in that because of their small size for a given output they can be largely factory built, and expensive 'on-site' work is less than for larger reactors, an important point when dealing with non-industrialised countries. It is, however, the small size of these reactors and their consequent possible safety problems which has led to widespread public opposition to the rapid increase in the nuclear programme, and this opposition has tended to spread to other reactor types. Of these the A.G.R. development has been confined to the U.K., but the Canadian persistence with the Candu system has led to its adoption in limited numbers outside its country of origin. Reactor development to commercial station size is expensive and development of other reactor types has been inhibited by the depressed economic position in recent years. In particular the development of the H.T.R. which has considerable potential advantages, not only for highly efficient electricity generation but also as an industrial heat source, awaits international co-operation. Western Germany and Japan are the main countries involved.

It is doubtful, however, whether extensive further work on slightly enriched reactors is now really justified, when it is generally recognised, outside the United States, that the long-term future of nuclear power depends essentially on the development of the fast breeders to which considerable effort is being devoted in most advanced countries.

The total installed nuclear generating capacity in 1978 is approximately 125,000 MW(e) with about half of this in the United States, followed by Japan, Western Germany and the United Kingdom with about 10,000 MW(e) each and then France, Russia and Sweden. Predictions are reasonably firm for a global nuclear generating capacity of about 500,000 MW(e) by 1985, since most of this additional capacity is under construction or firmly ordered. The U.S.A., Western Germany, France and Japan show the most rapid increases during this period (Fig. 33).

The probable nuclear capacity beyond 1985 is more speculative. In 1972 the United States Atomic Energy Commission (U.S.A.E.C.) and the European Nuclear Energy Agency (E.N.E.A.) made forward projections to 1990, which were in good agreement, each having a lower limit of about 0.8 million MW(e) and an upper limit of about 1.3 million MW(e). The median estimate could be smoothly extrapolated to 2.0 million MW(e) by the year 2000, and this was in good agreement with the 1973 O.E.C.D. estimate of between 2.0 million MW(e) and 2.5 million MW(e) by the end of the century. These estimates are now generally regarded as over-ambitious and more recent estimates for the O.E.C.D. areas anticipate between 0.83 million MW(e) and 1.64 million MW(e) by the year 2000, which will then represent some 40% of the world electricity-generating capacity. These projections are shown in Fig. 36.

The reduction in the nuclear generating capacity targets has been caused to some extent by the depressed economic situation in recent years leading to a downward revision of estimates for future total electricity demand. Delays in the P.W.R. programme due to public concern on safety has also played a part. Whether this concern is justified or not the reduction in the slightly enriched reactor programme will help to avoid the premature depletion of fissile reserves discussed in Chapter 6.

Present projections, however, indicate that the global reactor programme to the year 2000 will still be dominated by the light-water-moderated reactor with perhaps 7 to 8% of Candu reactors and a smaller percentage of graphite-moderated gas-cooled reactors with the H.T.R. possibly contributing to this in the 1990s. By the year 2000 about 10% of commercial nuclear plant should be of the fast-breeder type. A smaller proportion of slightly enriched poor conversion ratio reactors and more Candu, Magnox and H.T.R. reactors would represent a much healthier mixture of reactor types for the long term, with, of course, the introduction of the fast breeders as soon as plutonium stocks permit.

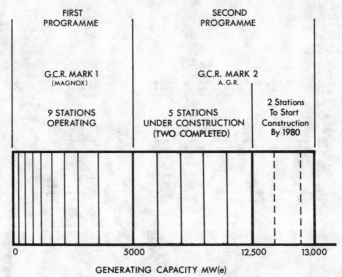

FIG. 31. The immediate U.K. reactor programme.

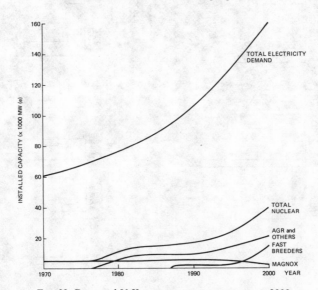

FIG. 32. Proposed U.K. reactor programme to A.D. 2000.

98 National Nuclear Power Programmes

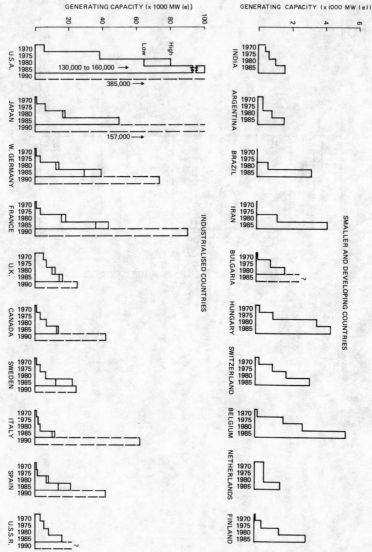

FIG. 33. Actual and planned national nuclear generating capacities.

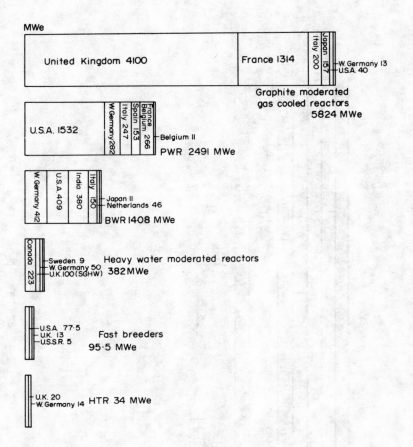

Fig. 34. Nuclear generating capacity by reactor type (1969).

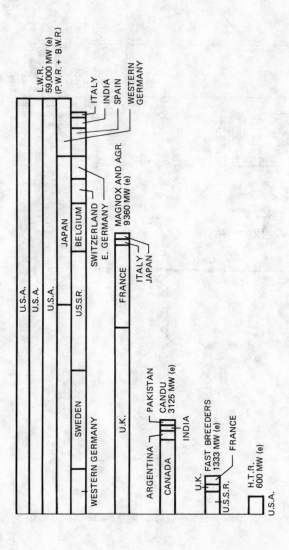

Fig. 35. Nuclear generating capacity by reactor type (1979).

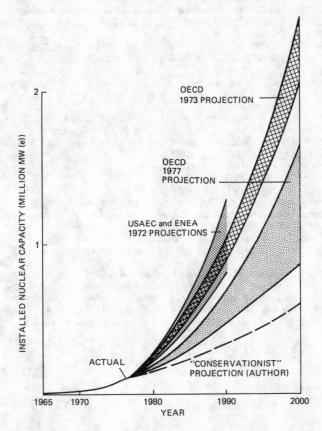

FIG. 36. Forward projections of nuclear power installations to A.D. 2000 (non-Communist world).

References

AVERY, D. G., BOGAARDT, M., JELINCK-FINK, P. and PARRY, J. V. L. (1971) Centrifugal Plants in Europe. Fourth International Conference on Peaceful Uses of Atomic Energy, Sept. 1971.
HEINRICH, D. O. (1978) Nuclear Energy in the Federal Republic of Germany. *J. Br. Nucl. Energy Soc.* **17**, no. 3, July 1978, pp. 163 – 169.

LARSON, CLARENCE E. (1970) The Present State and Future Outlook of Nuclear Power Generation in the United States. U.S. Atomic Energy Commission Tokyo, 1970 (A.E.A. Publicity Release).

McTIGHE, E. P. (1970) Helsinki, October 1970. Reported in *Atom*, Dec. 1970.

MOHRHAUER, H. (1972) Enriching Europe with the Gas Centrifuge. *New Scientist*, 5 Oct. 1972.

Nuclear Power—Its Development in the United Kingdom, R. F. Pocock, 1977. Unwin Brothers and the Institution of Nuclear Engineers.

O.E.C.D. Press Release 9 Mar. 1976.

PECQUENT, M. (1975) French Programme in the Area of Reactors and Enrichment of Uranium. *J. Br. Nucl. Energy Soc.* **14**, no. 1, Jan. 1975, pp. 11 – 16.

Proceedings of the International Conference on Nuclear Power, Salzburg, 1977.
    Paper IAEA-CN-36/581, V. Baun.
    Paper IAEA-CN-36/92, W. J. Schmidt-Küster.
    Paper IAEA-CN-36/217, M. Boiteux.
    Paper IAEA-CN-36/397, R. W. Fri.
    Paper IAEA-CN-36/53, S. Catchpole and F. P. Jenkins.
    Paper IAEA-CN-36/396, E. A. Wiggins.
    Paper IAEA-CN-36/492, J. Miida, W. Haussermann and S. Mankin.
    Paper IAEA-CN-36/179, J. S. Foster and S. H. Russell.
    Paper IAEA-CN-36/219, G. Vendyres, G. Besse, M. Rosonhole and B. Saitcevsky.
    Paper IAEA-CN- 36/356, O. D. Kazachkovskij.

SLIZOV, V. The Research Programme in Nuclear Energetics at Minsk Nuclear Power Institute, U.S.S.R. Paper read at Birmingham University, 4 Feb. 1975.

The Breeder Reactor. Proceedings of a Meeting at the University of Strathclyde, March 1977. Editor J. S. Forrest, F.R.S., Scottish Academic Press.

CHAPTER 10

Nuclear Power and the Environment

THE possibility of reactor accidents and their effects have been discussed in a previous chapter. It is also appropriate to examine the possible deterioration in the environment which could result from the normal operation of reactor power stations on the increased scale which will be necessary if they are to contribute substantially to meeting our future electricity requirements.

In considering the impact of any innovation on environmental conditions we almost instinctively postulate the intrusion of a new and sinister factor into a previously idyllic situation. In the developed countries environmental conditions started to be affected by industrial pollution several generations ago and it should be appreciated that pastoral uncontaminated living conditions are in some degree basically incompatible with high population densities and high material standards of living. We must therefore consider the pollution from the growth of nuclear power as an alternative to that from the growth of the equivalent electricity-generating capacity from coal- or oil-fired stations.

The London 'smog' of 1952 caused by fossil fuel combustion products was responsible for between 3500 and 4000 deaths. Subsequent 'Clean Air' regulations in the United Kingdom and in many other industrial countries have reduced but not eliminated this hazard. In the United States, for example, the 1970 'Clean Air Amendment' led to a reduction in the sulphur-dioxide discharge from 27.0 million tons in that year to 24.3 million tons in 1974, and miscellaneous particulate matter dropped from 8.3 million tons to 5.9 million tons in the same period. There continues to be a strong correlation between the death rate from 'bronchitis' and airborne sulphur-dioxide levels. These and other combustion products also cause considerable damage to buildings while increasing levels of CO_2 could have undesirable long-term climatic effects.

Not all the pollution from conventional stations is airborne. Slag heaps from coal mines are a potential danger as well as a visual blemish and the increased use of oil for power generation has led to the contamination of almost all the beaches in Western Europe. Some of these sources of pollution could be avoided by a more responsible attitude by the bodies concerned, but most of the airborne pollution could only be appreciably reduced by adopting costly measures which would substantially affect the economics of the conventional power stations. Perhaps one of the main dangers of 'conventional' pollution is that it has grown gradually with increasing population density and improved living standards and until very recently we have tended to accept this familiar but increasing hazard unquestioningly.

The same cannot be said of possible pollution from nuclear-power stations. The main hazard, radiation, was first observed in the form of X-rays by Röntgen in 1895 and radiation from naturally occurring radioactive isotopes was observed by Becquerel a year later. Several early radiation workers paid the penalty of uncontrolled exposure to radiation, and in so doing assisted us to gradually formulate safer levels of exposure and working procedures. They were, however, few in number and, apart from medical applications, the impact of radiation on the everyday lives of the general population was minimal until the discovery of fission.

The layman's introduction to the nuclear age was sudden, fierce and brutal: the dropping of nuclear weapons on Hiroshima and Nagasaki with 105,000 dead and 125,000 injured, many of whom were to die later due to radiation-induced disease. No major scientific discovery can have had a more disastrous introduction and the lay mind has understandably remained suspicious of any development involving the nucleus and radiation.

Radioactive material represents two possible biological hazards: as a source of α-, β- or γ-rays it can subject the body to these radiations from outside, or it may be inhaled or swallowed, subjecting the body to irradiation from within.

The metabolism of the body is such that certain elements are concentrated in specific organs and the radioactive isotopes of these elements behave in a similar manner. For example, radioactive iodine, ^{131}I, is concentrated in the thyroid, radioactive caesium, ^{137}Cs, concentrates in the muscle, strontium, ^{90}Sr, in the bones, whilst others such as xenon, Xe^{135}, are distributed more widely throughout the body tissue.

The time for which these isotopes are harmful within the body depends on the half-life of the radioactive decay proper of the isotope and on the metabolism of the element within the body. Isotopes with short half-lives will cease to represent a hazard due to the rapid reduction in their radioactivity, even if they remain chemically fixed within the body for an appreciable time. Others with long half-lives may be quickly eliminated because of the rapid metabolism for that particular element. Clearly isotopes with long half-lives which are also retained in the body for an appreciable period represent the worst hazard. Maximum permissible body burdens have been calculated for each radioactive isotope after consideration of both the radioactive half-life and the metabolic rate. These are the amounts of the isotope which the human body can contain without any discernible clinical ill effects. The maximum permissible body burdens for the most significant isotopes produced in reactors is shown in Table 5. These are expressed in terms of microcuries, a microcurie being approximately equal to the activity of a millionth of a gram of radium.

It is often possible to reduce the body burden of a specific radioactive isotope by taking large doses of the non-radioactive element. This then dilutes or displaces the radioactive isotope and much of the latter is then eliminated from the body. For example, in the case of excessive ingestion of ^{131}I, the consequent concentration of this isotope in the thyroid can be very appreciably reduced by taking large doses of normal iodine.

Whether the radiation originates in a source outside or inside the body, its effect is to ionise the body tissue, that is to remove electrons from the atoms making up body tissue, and the units in which this damage is measured are called rems. α-particles produce intense ionisation over a very short range, typically of the order of a tenth of a millimetre, γ-rays produce less intense damage over a very much larger range, of the order of a metre, and β-particles are intermediate between α- and γ-rays both in terms of intensity of tissue damage and range.

TABLE 5. *Maximum Body Burdens for Some Fission Product
Nuclei and Reactor Fuels*

Isotope	Critical organ	Maximum body burden (microcuries)
^{90}Sr strontium	Bone	1
^{91}Y yttrium	Bone	3
	G.I.	0.02
^{95}Nb niobium	Bone	44
^{99}Mo molybdenum	G.I.	0.01
^{96}Tc technetium	G.I.	0.3
^{103}Pd palladium	Kidneys	7
	G.I.	4
^{109}Cd cadmium	Liver	45
^{131}I iodine	Thyroid	0.6
^{133}Xe xenon	Total body	320
^{135}Xe xenon	Total body	100
^{137}Cs caesium	Muscle	98
^{140}Ba barium	Bone	1
	G.I.	0.7
^{140}La lanthanum	G.I.	0.9
^{144}Ce cerium	Bone	1
^{143}Pr praseodymium	G.I.	3
^{147}Pm promethium	Bone	25
	G.I.	0.3
^{151}Sm samarium	Bone	90
	G.I.	14
Natural uranium (soluble)	Kidney	0.04
	G.I.	10^{-3}
Natural uranium (insoluble)	Lungs	0.01
^{239}Pu plutonium (soluble)	Bone	0.04
	G.I.	0.02
^{239}Pu plutonium (insoluble)	Lungs	0.02

G.I., gastro-intestinal tract.

Yet biological damage due to radiation is not a phenomenon
specific to the nuclear age. Men and animals have been irradiated both
from outside and from within the body since the beginning of time. At
sea level we all receive some 30 millirems (30 thousandths of a rem) per
year of radiation from cosmic rays, and appreciably higher doses at
high altitudes. The slight radioactivity of indigenous rocks contributes
another 50 – 100 millirems per year depending on the locality, so the
'natural' level of externally received radiation varies from about 80 to

150 millirems per year. In addition to this our own bodies contain potassium and a small fraction of this is the naturally occurring radioactive isotope ^{40}K. We are subjected to approximately 20 millirems per year of internal radiation from this source.

The effect of radiation on the body at this 'natural' level is difficult to estimate. We have no radiation-free species with which to compare ourselves, but the natural levels vary by about a factor of 2 in this country and reach appreciably higher levels elsewhere, due either to higher levels of radioactivity of indigenous rocks or higher cosmic-ray levels at higher altitudes. No significant clinical differences have been noted amongst genetically similar groups living in different levels of natural radioactive background, and one therefore concludes that 'natural' levels of radiation of, say, up to 200 millirems per year are clinically harmless, or at least that the effect is so small as to be undetectable. At this level, however, there is strong evidence that radiation has genetic effects. An appreciable fraction of mutations is produced by the action of radiation on the parents. Some mutations are more viable than the original species, and if this is the case they eventually dominate as a result of natural selection. In the early part of the earth's history natural radiation levels were higher and mutations correspondingly more frequent; they were probably responsible for the evolution of man and the higher animals and even now may be responsible for the appearance of the occasional genius. As a conservative general rule, however, it is safer to assume that all mutants are inferior to the original species and consequently that all radiation is harmful and should be reduced to the lowest possible level. There is no 'safe' level of radiation, just as there is no 'safe' level of the gaseous oxides of sulphur or nitrogen.

The remaining question is how much radiation should we be prepared to tolerate in order to have adequate supplies of electricity and these higher material living standards which it brings?

At the natural level, say 100 – 200 millirems/year, radiation may produce a small number of mutations in future generations. Some may be superior, but the majority are probably inferior in some possibly minor aspect to the parent species. We have no option but to tolerate this.

We know from animal experiments and have, regrettably, evidence from nuclear weapons that a dose of 400 rems delivered over the whole body in a short time will kill some 50% of the exposed population; 400 rems to the whole body is the so-called median lethal dose. How do we set acceptable levels between these two extremes? Recommendations on this question are made by an international body, the I.C.R.P. (International Commission on Radiological Protection). In adopting the basic principle that all radiation is harmful and must be reduced to the minimum practical level, they recommend that members of the general population should not be exposed to 'artificial' radiation doses which are significantly higher than 'natural' radiation levels. An upper limit to the dose rate to which a member of the general public may be exposed is 0.25 millirem per hour and he would have to be so exposed for about 500 hours per year for the 'artificial' dose to exceed the natural dose. The direct radiation level from an operating reactor is less than this at any point on or near the reactor site to which members of the general population have uncontrolled access.

In the U.K. the effect of nuclear power stations on their immediate environment is the subject of open discussion between liaison committees consisting of C.E.G.B. staff, local residents and their expert advisers. Environmental measurements taken independently by C.E.G.B. and the Ministry of Agriculture are made freely available at these discussions.

The maximum dose permitted for personnel actually employed in the nuclear industry is ten times that for the general population and no worker must accumulate a dose of more than 5 rems each year. The doses are carefully monitored and the so-called 'classified workers' are subjected to frequent medical examinations and blood tests. Over the twenty years for which significant numbers of people have been exposed to this higher dose rate there has been no evidence of physical deterioration attributable to radiation in spite of predictions that a significant increase in leukaemia and possibly other forms of cancer could be expected. This situation compares very favourably indeed with that in the coal mines, where a very large proportion of the workers contract respiratory diseases within a twenty-year period.

So far as the general population is concerned, the dangers due to direct radiation from normally operating nuclear reactors is clearly minimal. The potential hazard represented by the accumulation of radioactive fission products in the normal operation of the reactor programme cannot be dismissed so lightly.

Fission products build up in the reactor fuel elements during the operation of the reactor and after some 40% or so utilisation of the fissile material the fuel elements must be removed from the reactor and treated to separate the valuable $^{239}_{94}Pu$ 'bred' in the reactor and the residual uranium from the fission products. In this country this operation is carried out at Windscale in Cumberland for all C.E.G.B. reactors and the highly radioactive 'spent' fuel elements are transported to Windscale in thick lead containers or 'coffins' which have the double function of attenuating the radiation to an acceptable level on the outside of the 'coffin' and protecting the spent fuel elements from damage and consequent escape of radioactive material in the event of accidents. The coffins are designed to withstand a 30-ft drop and temperatures up to 800°C without damage.

Chemical separation of the spent fuel elements is done by remotely controlled processes in heavily shielded buildings and the degree of extraction of fission products is such that 99.98% of all the radioactive waste is extracted in a form suitable for concentration in liquid form to an activity of about 10^4 curies per gallon. Some of the radioactive fission products have short half-lives so that the activity falls initially quite rapidly, but many have half-lives measured in years, and even tens and hundreds of years. The concentrated fission product liquid is stored in water-cooled stainless-steel containers of sufficient thickness to attenuate the radiation at the outer surface to acceptable levels and to resist corrosion for an estimated minimum period of 100 years. These containers are at present stored at ground level under constant supervision.

Prolonged surface storage is obviously undesirable but after ten years or so the heat output has fallen to a level where fission products can be vitrified into a solid mass for long-term or ultimate storage. Vitrification, of course, removes the necessity of relying on the integrity of an outer container since, although still contained in steel cans, the vitrified waste would itself be proof against leaching out of radioactive material should it come in contact with water at some future time. In the U.K. plans are well advanced for storing this vitrified waste in underground chambers, some 300 metres deep in stable rock formations. These will then be sealed to prevent access by future generations. The use of stable clay and salt deposits is being considered in other countries (Fig. 37).

Unattractive as this concept may appear it is necessary not to exaggerate the scale of the problem. If all the electricity in the U.K. were produced by nuclear power the yearly accumulation of fission product waste per head of the population could be contained in a vitrified capsule the size of an aspirin tablet, or, say, about 10 tons of vitrified waste annually for the whole country. After a thousand years the radioactivity of the waste will have decayed to that of the uranium ore from which it originated, and which, of course, is deposited by nature in areas freely accessible to man (Fig. 38). The main cause for concern, in the author's view, is not the 99.98% of the fission product waste to be treated in this way but the remaining 0.02%. In the United Kingdom this is at present diluted and discharged as relatively low-level liquid waste through a 2-mile-long pipe into the sea off Windscale. Extensive monitoring of the neighbouring coastline is being carried out and this includes measurement of the levels of radioactivity in seaweed and fishes, both of which accumulate radioactive materials to which they have been exposed.

The reprocessing of spent fuel elements at Windscale has recently been the subject of a public enquiry under the Chairmanship of Mr. Justice Parker. The Enquiry which lasted a hundred days was prompted not by any malfunction of the present plant but by an application of the plant operators British Nuclear Fuels Ltd. to build a larger plant to process the more active oxide fuel arising from the A.G.R. programme and from slightly enriched foreign reactors.

Although its terms of reference were specifically to look into the acceptability of the larger fuel-processing plant, the Windscale Enquiry was generally regarded as being of much wider significance. To many it seemed that the future of nuclear power as such was on trial, in the U.K. at least. Environmentalist groups such as 'The Friends of the Earth' were fully represented and the Windscale Enquiry seemed to meet the requirement of a public debate which the previous indeterminate 'Flowers Report' had called for. Justice Parker's findings were far from indeterminate, and gave a firm endorsement for the construction of the new facility, although calling for a tighter degree of control on radioactive effluent levels.

The proposed fast-breeder programme was not specifically included in the terms of reference of the Windscale Enquiry, but discussions on the plutonium fuel cycle were inevitable, and indeed arose naturally from the basic question of whether to process spent fuel, thus segregating the plutonium, or to store it in an unprocessed state. On this specific point the Windscale Enquiry was quite clear that it was in the best interest of future generations to process the spent fuel elements as soon as practicable, thus removing the plutonium and some other long-lived actinides before final storage by the method described above.

There is, of course, a long-term economic incentive to extract the residual ^{235}U, the ^{238}U and the plutonium from spent fuel elements for use in future reactors, and in particular for the fast breeders for reasons discussed in Chapter 7, and it is this recycling of plutonium the so-called 'plutonium economy', which has been a major source of debate in many countries in recent years.

It is now fairly generally conceded that fast-breeder reactors, which are an essential component of the plutonium economy, can be made acceptably safe (Chapter 6) and the main causes of concern can be summarised under three headings: that plutonium is so toxic that its processing involves an unacceptable risk to the workers and an accidental release would represent an unacceptable risk to the community generally; that as a weapons material plutonium would be an attractive target for terrorists and that the widespread availability of plutonium from commercial reactors would greatly increase the risk of nuclear war. These objections clearly deserve detailed consideration.

Plutonium is indeed a toxic material, but not more so than many substances currently used in industry. The main danger is due to the inhalation of insoluble particles which could then form local spots of intense irradiation in the lung, with increased possibility of lung cancer. Great precautions are taken in the handling of plutonium and classified workers both in the United Kingdom Atomic Energy Authority and in British Fuels have shown a lower incidence of lung cancer than that expected on statistical grounds from a normal population of the same age distribution. There have, of course, been instances of death through lung cancer and it is impossible to prove that some of these were not due to plutonium inhalation, but the risk appears to be much less than for many accepted industrial processes. It is clear that the utmost care must continue to be taken in the handling of plutonium and safety levels must be under constant review. Nevertheless it is necessary to preserve a more balanced attitude towards this potential hazard which has been grossly magnified in the popular press. For example, several workers in the Atomic Weapons Research Establishment have recently been reported to have accumulated twice the maximum permissible body burden of plutonium, a level which would represent an added lung cancer risk comparable with that from smoking one cigarette per day, yet this attracted more publicity than did the deaths of over a hundred holidaymakers due to a crashed petrol tanker in Spain at about the same time.

The probability of a release of plutonium from a processing plant to the general environment is low, and certainly less than the risk of release if untreated fuel elements were stored over long periods as advocated by some environmentalist groups.

The danger that terrorists could hi-jack plutonium for the purpose of political blackmail figures largely in the 'Flowers Report'. Any such danger could clearly be minimised by carrying out spent-fuel processing and fabrication on the new fuel elements on the same site, thus avoiding the necessity of transporting plutonium in a highly concentrated form.

The highly radioactive nature of the spent fuel elements would make them more of a danger to the hi-jackers than to any intended victim should they attempt to remove them from the transport containers, though in the containers they are, by design, harmless. Newly fabricated fast-breeder fuel elements contain about 20% of plutonium, and at this concentration they would be useless for weapon construction. They are not themselves highly radioactive but could easily be made so if this were judged to be necessary to deter possible terrorists.

Existing chemical plant, 'liquid-gas' containers in transit and even petrol tankers would make a much more suitable target for hi-jackers than would either new or spent fuel elements.

Any fuel-element treatment and fabrication site would obviously need to be well guarded and this is seen by some as posing a threat to civil liberties and democratic institutions, but it is difficult to accept the logic of this argument when many other installations of importance to national security are already under armed guard.

The possible proliferation of nuclear weapons is frequently associated with the 'plutonium economy' required for commerical fast breeders. It should be remembered, however, that the major powers now hold sufficient stocks of enriched fissile material to annihilate most of humanity, and none of this was produced from commercial fast breeders. It came largely from military reactors purpose built to produce pure 'weapon grade' plutonium and from the ^{235}U isotope separation plants. A nation determined to produce nuclear-weapons material could do so much more effectively by either of these methods than by building a commercial fast breeder, which would be much more expensive and produce plutonium containing a high proportion of heavier isotopes and consequently much less suitable for weapons manufacture.

It is true that a widespread fast-breeder programme would lead to the production of appreciably more plutonium than in the case of a thermal reactor programme of the same generating capacity, but in the fast-breeder case the plutonium becomes an essential and consumable part of the fuel cycle, and the amount in stock at any one time would be far less than the accumulated 'surplus' plutonium stocks from a thermal 'non-breeder' programme.

The answer to any increased danger of proliferation of nuclear weapons introduced by commercial fast breeders is seen to lie in the control of the spent-fuel processing and fuel-fabrication plants. The existing Non-Proliferation Treaty already provides for this control in that the six existing nuclear nations have agreed to provide nuclear fuel for those not having fuel-processing facilities in exchange for the assurance that they will not themselves construct such plant. This 'new fuel for old' agreement places the large powers in a strong position to prevent the misuse of plutonium produced in commercial reactors, but a refusal to honour it on their part would encourage smaller nations to set up their own facilities.

For reasons already outlined, however, a commercial fast-breeder programme would do little to affect the existing threats of nuclear war, which really started with the dropping of the first nuclear weapon at Hiroshima. It could, of course, be argued that the possession of nuclear weapons by the great powers and the universal recoil from the horror of their use has maintained the uneasy world peace for the past thirty years. It has at least limited the conflict to clearly defined geographical areas and, by common agreement, to the use of conventional weapons. The main hope for mankind may well lie in the fact that this healthy instinct for self-preservation which has guided the policy of the major powers will be no less marked in the smaller nations, if and when they acquire nuclear weapon potential, as they quite clearly can by methods which are much easier than the building of commercial fast-breeder stations.

National attitudes to the fast-breeder reactor and its attendant plutonium economy appear to have been guided in no small measure by their present energy requirements. The United States, with a high proportion of the world's fossil fuel reserve and relatively abundant fissile fuel has, for the time being, decided against fuel reprocessing in view of the alleged dangers of the plutonium fuel cycle. This may appear to be a surprising stand to take by a nation which has produced more than a hundred times more plutonium in its weapons programme than from its civil nuclear reactors to date. It may, perhaps cynically, be attributed to commercial pressures to export the P.W.R. reactors with their possibility of a 'break-even' thorium breeder core, though there is, of course, no evidence that ^{233}U would present a lesser hazard than plutonium. The French with their dwindling fossil fuel reserve and limited fissile reserves are the most advanced in their acceptance of the fast breeder, and their 'Super Phénix' 1300-MW(e) fast breeder has obtained official planning approval. In the U.K., coal reserves are somewhat higher and there is an over-provision of electricity-generating at present, so that more protracted discussions are still going on regarding our own proposed 1300-MW(e) C.D.F.R.I. which cannot be authorised without yet another public enquiry.

The Fast Breeder debate apart, however, recent investigations indicate that the environmental health hazard associated with the nuclear-power programme are considerably less than those associated with 'conventional' coal or oil generating plant. It is perhaps not strictly relevant to discuss the environmental impact of oil-fired stations, since they must be regarded as a temporary phase, but such incidents as the *Torry Canyon* and the *Amoco Cadiz* have served only to highlight a pollution menace which has been growing steadily over the past twenty years.

The airborne effluent from fossil-burning stations has already been referred to, and in passing it may be of interest to note that the actual output of airborne radioactive material from a coal-burning station is greater than that from a Magnox station of comparable power. In addition to this, however, one must consider the health hazard involved in the coal-mining industry and the environmental damage of the pit-head waste. Approximately sixty miners per year are killed in underground accidents in the United Kingdom and a high proportion are disabled by pneumoconiosis by middle age. It is possible to deduce with some accuracy the cost in human life of coal-fired stations, and the Ford Foundation Report on American plant puts this at two deaths for each year's operation of a 1000-MW(e) station from causes other than air pollution. These are mainly connected with coal mining and transport. The premature deaths due to respiratory conditions aggravated by gaseous effluent are put at between 18 and a staggering 50 deaths per year per 1000 MW(e) for existing stations depending on location and the sulphur content of the coal. New stations built to conform to the Clean Air Amendment should reduce this to between 0.4 and 25 deaths per year.

Against this the cost in life of a 1000-MW(e) nuclear station is estimated to be between 0.6 and 1.0 death per year, and this includes workers in uranium mining and milling (0.1 – 0.25), accidents in reactor construction work (0.1 – 0.25), exposure to radiation in reactor operation and fuel reprocessing (0.2 – 0.3)* and the general public due to increased radiation levels (0.2).

The possibility of reactor accidents has been calculated statistically to add a further 0.02 possible fatalities per reactor year (WASH 1400), although this is regarded as an under-estimate in the case of P.W.R.

---

*The health record of radiation workers in the U.K. (Sir John Hill, 'The Abuse of Nuclear Power', *Atom* **239** (1976), p. 236) would not support this estimate. Radiation workers had a better health record than that of the general population in the period 1962/74.

reactors in the Ford Foundation Report. They also calculate that the effect of fast breeders and the attendant plutonium cycle would reduce the fatality estimate due to uranium mining and milling by 0.04 per reactor year and increase those of the general public by 0.07 per reactor year.

Whilst quantitative results from such statistical analyses must be treated with caution the margin in favour of nuclear rather than coal-fired stations appears to be very substantial indeed and this is borne out by the actual record in the U.K. (Table 6). Whilst there has been a most welcome fall in coal-mining accidents in recent years, these still far exceed those in the nuclear-power industry.

TABLE 6. *Comparable Occupational Hazards of Coal and Nuclear Programmes (U.K.)*

Accidents per 100 workers (including over-exposure to radiation)

	Nuclear-power industry			Coal-mining industry		
	Fatal	Serious	Non-serious	Fatal	Serious	Non-serious
1965	Nil	Nil	0.1	0.05	0.24	42.3
1970	Nil	Nil	0.06	0.03	0.21	29.5
1973	Nil	Nil	0.12	0.03	0.21	24.3
1976	Nil	Nil	0.06	0.02	0.21	19.6

It is tempting to assume that renewable energy sources are totally benign from an environmental standpoint, yet this is not the case. The sighting of large windmills, if large-scale wind power ever does become technically feasible, will produce an obvious aesthetic problem while its obvious effects on drainage and sewage disposal have been a major factor in inhibiting the Severn estuary tidal power scheme. Even solar energy is not completely exonerated. The accident rate in the home is approximately equal to that on the roads and a high fraction of these are produced by falling from ladders. A simple statistical calculation will show that if they were universally adopted, the probable casualties in cleaning domestic solar-heating panels would far exceed those from the nuclear-power programme.

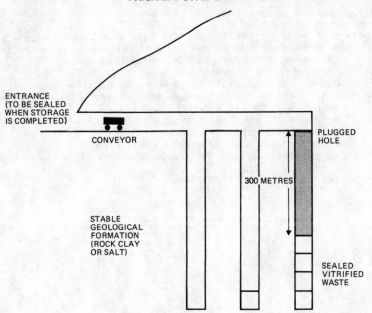

FIG. 37. Disposal of vitrified radioactive waste.

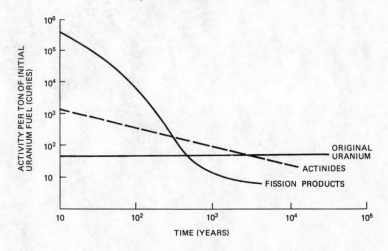

FIG. 38. Decay of radioactive products from fission reactor.

References

CLELLAND, D. W. (1977) 'The Management of Radioactive Waste for Reprocessing Operations', *The Breeder Reactor*—Proceedings of a Meeting at the University of Strathclyde, 25 Mar. 1977, Scottish Academic Press.

HILL, Sir JOHN (1976) 'The Abuse of Nuclear Power', *Atom* **239**, 236.

Nuclear Power Issues and Choice (1977) Ford Foundation Report. Bollinger.

SAGER, L. A. (1974) 'Health Costs Associated with the Mining, Transport and Combustion of Coal', *Nature* **250**, 197.

The Flowers Report (1976) Sixth Report of the Royal Commission on Environmental Pollution, H.M.S.O., Sept. 1976.

The Hazards of Conventional Sources of Energy (1978) H.M.S.O.

The Rasmussen Report WASH 1400 (1975) American Nuclear Regulatory Commission, Oct. 1975.

The Windscale Report (1977) H.M.S.O.

CHAPTER 11

Renewable Energy Sources

THE amount of solar heat received at the earth's surface is estimated to be about 5000 Q annually, or approximately 25,000 times greater than our present requirement for controlled energy for industrial and domestic use. This heat is, of course, responsible for maintaining the earth's surface at an acceptable temperature, for the growth of crops, or in short, for life as we know it, but it has proved surprisingly difficult to 'harness' even a small fraction of it for domestic and industrial purposes.

One of the reasons is that, despite the large total amount, the average heat flux on the earth's surface is only about 200 watts per square metre. This is to be compared to some 50 kilowatts per square metre which is the thermal flux in the heat exchangers of a commercial power-generating plant. To make efficient use of solar energy for electricity production using a conventional steam cycle, therefore, it would need to be concentrated by a factor of 250 or so. This, of course, could be done by placing boilers at or near the foci of large concave mirrors and focusing direct sunlight in order to produce steam and generate electricity in the normal manner. The idea of focusing the sun's rays for heating is not new; primitive cooking facilities employing concave mirrors have long been used in tropical countries, and the Russians are reported to have attempted to construct a solar power station along the lines suggested as early as 1961 (Allibone). Little has been heard of it subsequently, presumably due to the limited amount of direct sunlight in northern climates.

It appears at least possible, however, that a ring of concave mirrors constructed around the equator focusing the sun's rays on suitable water boilers could be technically feasible. A mirror area of a few thousand square kilometres would be adequate to meet the world's present electricity demand but the problems of long-distance electricity transmission would be considerable even if obvious political problems could be overcome.

In more temperate climates direct solar energy is ill matched to our industrial and domestic energy demand, reaching its maximum in summer when the energy demand is at its lowest, and also having a daily variation which does not coincide with that of the energy demand. It is spasmodic due to frequent and unpredictable cloud cover, so that the possibility of a continuous and reliable energy supply appears remote. It has been suggested that the problem of cloud cover could be overcome by the use of satellites in which solar energy would be converted into microwaves which could be beamed down through the clouds to fixed locations on the earth's surface, but it is hard to suppose that any such scheme could be commercially viable.

Solar energy is already widely used for the direct production of small quantities of electricity using solar cells, and these are useful for highly specialised applications such as the powering of satellite instrumentation and radiobeacons in remote areas, but their efficiency is low and their capital cost too high by a factor of a hundred or so to permit their use for large-scale electricity production under present circumstances.

It is generally agreed that the most promising direct application of solar energy within the foreseeable future is in domestic heating and air conditioning. As early as 1958 Landsell reported that a house in Denver, Colorado, was maintained at a constant temperature of 90°F by using solar heat with a suitable storage system, and he predicted that by 1978 some 13 million homes in the U.S.A. would be heated in this way. This prediction is far from being realised and it is true to say that large-scale research to harness solar energy, even for domestic heating, only really started with the oil crisis in 1974.

The now 'conventional' domestic solar heating system is shown in Fig. 39. This employs a metallic 'solar panel' mounted on the roof and suitably orientated to receive maximum sunlight, shielded by glass outside to reduce reradiation and thermally lagged on the underside. Heat collected by the panel is transferred to the rest of the house by a water-circulation system. A 100-m² collector would be adequate to provide some 75% of the domestic heating and hot-water demand in a relatively sunny climate such as Southern Europe, but correspondingly less, perhaps 30% on average, in the U.K. In this non-focusing system some use can be made of the scattered sunlight through clouds, but in order to accommodate spells of cloudy weather the water-circulation system must have a large thermal capacity. About 10,000 to 15,000 litres are required for a normal home and clearly auxiliary heating must be provided to cope with the longer sunless periods in winter.

In the U.K. the system as detailed above would only be marginally economically viable under present conditions if initially planned in the building of a new house, where a capital investment of perhaps £1500 could reduce the fuel bill by 30%. Installation in an existing building would be more costly and consequently less attractive, but this situation could change with increasing fuel costs, and domestic solar heating and air conditioning is obviously more attractive in sunnier climates.

A recent proposal by Dr. W. E. J. Neal suggests heat collection by air circulation under the existing roof space coupled to a water system employing a heat pump and has the dual advantage of reducing installation costs and improving efficiency. Initial trials suggest that such a system may well appreciably improve the economics of domestic solar heating to be commercially attractive even in temperate climates.

The mismatch of solar heat with industrial and domestic energy demand would be less serious if a really economic method of energy storage could be devised, and this is an urgent requirement for most new energy sources, including nuclear power for reasons discussed in Chapter 8.

Several mechanical energy-storage systems have been proposed, including the storage of compressed gas in large natural caverns, but the most promising to date is the pumping of water from low-level sources to high-level reservoirs where its potential energy may be used to produce hydro-electric power as at Trawsfynydd nuclear-power station. The use of reversible endothermic chemical reactions is also being investigated, but these studies are still at a relatively early stage in terms of large-scale commercial exploitation.

A second way in which the mismatch could be overcome is the indirect use of solar energy in the growth of fuel crops. The use of wood as a fuel is, of course, almost as old as mankind, but the fuel value of the annual global wood growth at present is estimated at 0.05 Q, perhaps a quarter of the world's energy demand. It is, of course, an increasingly valuable structural material and the large-scale burning of existing wood supplies would be hopelessly uneconomic. A much more rapidly growing combustible material is required and several suggestions have been made such as the growth of the 'water hyacinth' in tropical regions and, perhaps more realistically, the wider growth of sugar cane where the cash value of the extracted sugar would supplement the economics of the waste-burning process. A 20-MW(e) plant of this type is planned at Kauai, Hawaii, for late 1980.

The use of solar energy to accelerate the decomposition of farm, domestic and industrial waste in order to produce combustible gases is another possibility in which the actual matching of the solar heat to the energy requirement is unimportant.

Yet the production of 'biomass' fuels and accelerated decomposition are not the only potential indirect uses of solar energy. The sun's heat and consequent thermal currents are responsible for the winds which sweep the earth's surface, and windmills were widely used as an energy source before the Industrial Revolution. Like solar energy itself, wind power is unpredictable and spasmodic in most areas, but in very general terms it is higher in winter when the energy demand is greatest. The actual energy density is relatively low and the picturesque windmills associated with the grinding of grain and the pumping of water for farm irrigation, once seen in large numbers in flat windy areas such as Holland and Norfolk, had a power output of only a few kilowatts. To obtain large powers the sails must be made appreciably larger, and this introduces mechanical strength problems. A 1.25-MW(e) installation built in Vermont, U.S.A., had 150-ft-long blades weighing 8 tons, and these proved to be too fragile for sustained operation in conditions of high wind. The use of blades mounted in vertical rather than the traditional horizontal axis reduces the problem of blade fracture and larger powers may be possible. At the peak of their use it is estimated that there were some 10,000 windmills in the U.K., and if this number of vertical-axis 1-MW(e) machines were built, they would produce about 10% of our total electricity demand. The economics of such a system would need to be investigated in some detail, as would its environmental impact since inevitably the installations would need to be sited on high ground often in areas of natural beauty. It seems unlikely that the use of large windmills on such a scale would be feasible, and consequently wind power is unlikely to make an appreciable contribution to our overall energy demand. Smaller installations could still have a useful role to play in the supply of small amounts of electricity to remote areas and in fulfilling their traditional tasks of milling and irrigation particularly in developing countries.

Winds passing over the oceans produce waves, which, since water is approximately 1000 times denser than air, have an appreciably higher energy density. This second indirect manifestation of solar energy is also spasmodic but less so than either direct sunshine or wind speed and it reaches an obvious overall maximum in winter months, thus roughly matching the energy demand. The U.K. is fortunate in this context in having regions of extremely turbulent ocean off the western coast of Scotland and in the south-west, and some 350 schemes for extracting energy from the waves have been patented in the past twenty years. Perhaps the most promising of these is the scheme proposed by Professor Salter of Edinburgh University. This consists of a row of floats, 'Salter's Ducks', mounted on a common axis and contoured in such a way as to convert the energy of the incoming waves into a racking motion, as shown in Fig. 40. A row of such 'ducks' rotating randomly about a common axis, which would, therefore, be approximately fixed in position, could, by a suitable combination of magnets and coils, be used to produce electricity for transmission to the shore. The feasibility of the scheme has been tested in water tanks and is now to be essayed using larger-scale models. The term 'duck' conjures up a rather misleading impression of size, since a full-sized float would be about the size of an ocean-going liner. It is estimated that a line of these some 400 km long off the Scottish coast could meet about 10% of the U.K. electricity demand. Capital costs would clearly be high, but 'fuel' costs zero and the environmental impact may possibly be beneficial in that the scheme would produce much calmer conditions in coastal waters. One disadvantage is that the region of electricity production would be remote from the main regions of demand. The economics of a large scheme of this type have yet to be assessed in detail, although early estimates are discouraging.

Although the sun is the major source of 'renewable' energy, the moon also has a part to play and perhaps, surprisingly, it is this in the form of tidal energy which has first been exploited on a large scale. The basic principle is simply to trap the incoming tide and to allow this to return at low tide via turbines to produce hydro-electricity, but in order to be economically feasible, a large tidal estuary is required. The Severn estuary is perhaps the most promising one in Europe and schemes for damming it to produce a large tidal hydro-electric scheme have been under consideration since the 1920s. The French Government have been somewhat more adventurous in damming the estuary of the Rance in Brittany. An 800-metre long dam produces 240 MW(e) and has been in operation for five or six years. Energy is available as the rising tides operate the turbines mounted in the dam wall and also at falling tide from the trapped head of tidal water and from the build up of river water, but no power is available when the river and tidal levels are equal. This period of zero output unfortunately varies from day to day according to the tidal timetable. The scheme was probably not economically viable when initially planned but has become so due to the rising costs of fossil fuels. Unfortunately, the number of suitable tidal estuaries available is limited, so that whilst such schemes may have local importance, their contribution to the global energy demand is likely to remain small. The same can be said of the more conventional hydro-electric schemes arising directly from the use of rainfall on high ground. Although hydro-electric schemes provide a high proportion of the electricity in mountainous countries such as Norway and Switzerland, the exploitation of all possible sites on a global basis would meet only 7% of our present energy demand.

Geothermal energy cannot strictly be regarded as a renewable energy source, since its exploitation would accelerate, albeit marginally, the cooling of the earth, the heat of which we could regard in this context as a capital asset. As we penetrate the earth's crust the temperature increases at a rate of about 30°C per kilometre, on average, though much higher thermal gradients exist in volcanic regions, where some heat reaches the surface in the form of pressurised hot water and steam escaping through faults in the earth's crust. These are already exploited for electricity production notably in California, New Zealand, Italy and Mexico, with smaller geothermal electricity-generating installations in Japan and the U.S.S.R. A world generating capacity of about 1880 million MW(e) is planned by 1980. Temperatures of about 250°C are required for reasonably efficient electricity generation and natural geysers of this type are limited in number. Naturally occurring hot springs at more modest temperatures are used for district heating and meet almost half of the domestic-heating requirement in Iceland. A more limited installation provides some district heating near Paris, but the hot springs in the U.K. have a maximum temperature of 48°C which is too low for even this use on a reasonable scale.

The utilisation of geothermal heat by pumping water into artificially produced craters some 5 km underground and extracting it at about 200°C is receiving serious consideration. Drilling to this depth is within the capabilities of oil-mining technology and it has been proposed that the underground crater could be produced by a nuclear explosion (project Ploughshares). It would probably be necessary to continually enlarge the crater by successive explosions because of the comparatively poor thermal conductivity of rock.

Clearly higher research expenditure on these newer forms of energy is fully justified and their development has been grossly neglected until very recently. Many are limited in total energy content, but may have an important part to play in geographically favoured locations or as an intermediate phase in the technology of developing countries. Others, notably solar energy, have tremendous total energy content but would be difficult to apply to the large-scale production of electricity on which sophisticated society is increasingly dependent.

In the U.K. research expenditure on renewable energy sources in 1977 was less than three million pounds, or about 2% of the research expenditure on nuclear power. This is to be increased to six million pounds, with 1.5 million allocated to further studies of a Severn estuary tidal scheme and 4.5 million divided between other potential renewable energy sources. Research expenditure in this field is also increasing rapidly in the United States and most Western European countries.

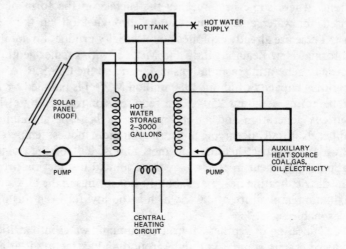

FIG. 39. A 'conventional' domestic solar heating system.

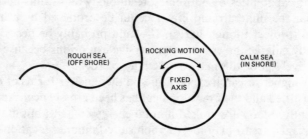

FIG. 40. 'Salter's Ducks.'

References

ALLIBONE, T. E. (1961) *The Release and Use of Atomic Energy*. Chapman and Hall.
DAWSON, J. K. (1976) Alternative Energy Sources for the United Kingdom. *Atom,* **231**, 11.
LANDSELL, N. (1958) *The Atom and Energy Revolution*. Penguin.
NEAL, W. E. J. (1978) Heat Pumps for Domestic Heating and Heat Conservation. *Phys. Technol.* **9**, 154.
SALTER, S. H. (1974) Wave Power. *Nature,* **249**, 720.

CHAPTER 12

The Fusion Programme

JUST as certain heavy elements ($^{235}_{92}U$, $^{233}_{92}U$ and $^{239}_{94}Pu$) undergo exothermic fission, the fusing together of light elements can be accompanied by the emission of energy. One of the most promising reactions is the fusion of the heavy isotopes of hydrogen, deuterium, according to the equation:

$$^2_1H \ + \ ^2_1H \begin{cases} ^3_1H \ + \ p \ + \ 4.0 \ \text{MeV} \\ \\ ^3_2He \ + \ n \ + \ 3.2 \ \text{MeV} \end{cases}$$

which produces either the still heavier isotope of hydrogen, tritium, plus a proton or the light isotope of helium, ^3_2He, plus a neutron. In each case the evolution of energy is considerable, approximately one part in a thousand of the total mass involved in the reaction being converted into energy; in other words as a method of converting mass into energy the above fusion processes have about the same efficiency as do the fission processes. The fusion of deuterium with tritium

$$^2_1H \ + \ ^3_1H \ \rightarrow \ ^4_2He \ + \ n \ + \ 17.6 \ \text{MeV}$$

is even more efficient in that almost four parts in a thousand of the initial mass is converted into energy. The possibility of obtaining energy from these reactions was postulated as early as 1929 (Atkinson and Houtemans, 1929) and hence predates interest in the fission process by ten years. Compared with the fission process, fusion has several important potential advantages. Deuterium is the heavy isotope of hydrogen and, although its isotopic abundance is only one part in 6000, reserves of deuterium in the form of 'heavy water' in the sea are appreciably greater than that of uranium or thorium, some 10^{10} Q or 100,000 times greater than the probable economically recoverable fission reserves, even assuming sensible use of the breeder reactions. The products of fusion processes are $^3_2$He, $^3_1$H and $^4_2$He, and of these the two helium isotopes are stable and therefore do not represent a possible radiation hazard. Tritium is radioactive but cannot in any sense be regarded as a waste product, since it undergoes fusion more efficiently than does naturally occurring deuterium. Any fusion reactor would therefore be designed to recover the tritium produced for economic reasons and with this any health hazard represented by $^3_1$H would be considerably reduced. A fusion reactor would therefore not produce radioactive waste, which, as discussed in Chapter 10, is one of the main residual problems in the fission reactor programme. A third possible advantage is that the total amount of radioactive material in a working fusion reactor is likely to be very much less than that in a fission reactor and in the event of an accidental release of reactor material the health hazard would be appreciably less. Tritium is the only directly produced radioactive material, and since it is an extremely light gas the hazard would be local. Intense neutron fluxes from a fusion reactor would render the constructional materials extremely radioactive (Steiner, 1971), but it is difficult to imagine an accident which would spread these over a wide area or reduce them to a condition in which they could be ingested. The incentive to produce a self-sustained controlled fusion reaction for electricity production is therefore considerable. Fusion processes formed the basis of the 'hydrogen bomb', illustrating that a fusion system can be produced to give a single uncontrolled release of energy. The reactions are also used in many nuclear laboratories as a source of neutrons and are produced in a controlled manner for this purpose by accelerating deuteron beams on to deuterium- or tritium-rich targets. Fusion reactions are thought to be the origin of the heat production on the sun, but it has not so far been possible to establish conditions on earth which will lead to a controlled series of self-sustained fusion reactions in such a way that the system as a whole has a positive energy release. When deuterons strike tritium or deuterium targets for neutron production the most frequently occurring interactions are ionising collisions between the incident deuterons and the 'target' deuterium or tritium. Most of the incident deuterons lose their energy in removing the orbiting electrons from the target nuclei and then have insufficient energy left to interact with the nucleus itself. The nuclear reactions, although individually having a positive energy output, are comparatively rare events and when one considers the energy necessary to run an accelerator to produce the accelerated deuterons, it becomes clear that an overall kinetic energy release cannot be obtained by using a simple accelerator system with solid or gaseous target.

As pointed out by Sir George Thomson in 1946, the most promising approach appears to be to supply enough energy to deuterium gas to produce first ionisation and then fusion. Ionisation is the dissociation of the deuteron, which is the nucleus of the deuterium atom, from its orbiting electron and at this stage one has instead of an atomic or molecular gas a system consisting of equal numbers of positively charged deuterons and electrons dissociated from each other and free to move independently. Such a system is called a plasma. The deuterons are mutually repulsive because of their positive charges and the next step is to give them sufficient kinetic energy to overcome the coulomb repulsive force and enable them to fuse together. One way in which this can be done is by heating the plasma. The thermal energy of the deuterons appears in the form of random vibrational motion and at a sufficiently high temperature this should have enough energy to overcome the coulomb repulsive forces. In principle this appears to be reasonably simple, but any vibrating charged particle emits energy in the form of electromagnetic radiation and this represents an energy loss to the system. The energy loss comes mainly via the electrons since these, being lighter, vibrate appreciably more rapidly than the deuterons, but it is nevertheless an energy loss from the plasma as a whole which increases with temperature. The number of fusions and consequently the energy output of the system also increases with temperature, and fortunately more rapidly than does the radiation loss. When the rate of energy production exceeds the rate of energy loss there will be an overall temperature increase in the system, and when this occurs we have a reaction produced by heat and in turn producing more heat giving the possibility of a self-sustained reaction, a so-called 'thermonuclear fusion reaction'.

Unfortunately the temperature at which heat generation starts to exceed heat loss, the 'break-even' temperature, is very high as shown in Fig. 41, approximately 200 million degrees Centigrade in the case of deuterium plasma and somewhat lower, about 30 million degrees, in the case of mixtures of deuterium and tritium.

These temperatures are well outside our previous experience and are compared with other high-temperature points in Fig. 42 (note that the temperature scale is exponential). Thus the required temperature is 20,000 times hotter than that at the surface of the sun.

In addition to the extremely high temperature requirement it is necessary that the fusing system should stay together for long enough to enable sufficient fusion processes to occur. We therefore have to obey an additional criterion, the so-called Lawson criterion, which says that in order to obtain an overall positive energy output the product of the deuteron density (number of deuterons/cubic centimetre) and the time for which the system is held together, the so-called 'containment time', must be greater than 2×10^{14} (cm^{-3} sec).

Methods by which these conditions could possibly be achieved and also the inherent difficulties are well illustrated by the now classic experiment of the Russian Kurchatov in the early 1950s. He enclosed deuterium gas in an insulating cylindrical vessel and placed electrodes at each end. The pressure of the deuterium was about 10^{-3} torr or about one-millionth of atmospheric pressure, a very easily attained pressure at which gas can fairly readily be ionised by an applied voltage. A moderately high voltage in the kilovolt range was applied across the electrodes and the gas became ionised and therefore conducted an electric current. The first heating mechanism was ohmic heating equal to I^2R where I is the current flowing and R the resistance of the plasma. This is the normal type of heating one observes in a domestic electric fire except that the conducting gas replaces the wire filament. It is in fact convenient for the purpose of explaining the second heating mechanism to consider the conducting gas as being made up of a large number of filaments each carrying a current. Each of these will produce a circular magnetic field in the plane at right angles to the current filament as shown in Fig. 43. The magnetic field due to a filament near the central axis of the discharge will therefore intersect the outer filaments at right angles and the interaction of field and current, according to Fleming's left-hand rule, will produce a force tending to attract the outer filaments towards the axis of the discharge.* Rather than filling the whole vessel, therefore, the conducting gas will be compressed or 'pinched' into a narrow axial filament (Fig. 43) and since the compression takes place rapidly it will result in a further heating of the gas, the so-called 'pinch-force-heating' mechanism. This is important for two reasons. First the ohmic heating mechanism is effective so long as the gas is not completely ionised, but when the gas reaches a temperature at which the electrons and deuterons are dissociated, that is when a true plasma is produced, the electrons particularly are free to move with very high velocities towards the anode or positive electrode and the resistance of the plasma falls to a very low value. This occurs at temperatures well below one million degrees so that a further heating mechanism is required to take over where ohmic heating stops and pinch-force heating is one such mechanism.

The second effect of the pinch force is even more important. If we consider the pinched axial plasma we see that it is actually held in place or 'contained' by its self-generated magnetic field (the so-called B_θ field). The excess pressure within the compressed gas is balanced by the inward pinch force resulting from the interacting of the current and the magnetic field ($F = B\Lambda i = \Delta P$). The fact that a plasma can be contained by a magnetic field rather than a material vessel is vital to the whole programme, since no known material would be capable of withstanding the required temperature of 200 million degrees.

Fusion research for the twenty years after Kurchatov can be summarised as the search for a magnetic field configuration which will hold a hot plasma in a stable manner at least for the time necessary to fulfil the Lawson criterion, and with the investigation of possible additional heating mechanism necessary to achieve the temperature requirements.

If Kurchatov's experiment illustrated the essential principles of plasma heating and containment it also illustrated some of the difficulties. Though the plasma may be prevented (somewhat imperfectly) from striking the vertical walls of the vessel it must inevitably come into contact with the metallic anode and cathode. Energetic electrons

---

*For those not familiar with the laws of electromagnetism it is sufficient to remember that currents flowing in the same direction attract each other.

striking the anode will eject less energetic ones, so the plasma is exchanging energetic or 'hot' electrons for less energetic or 'cold' ones, an additional energy-loss mechanism. The situation is even more serious at the cathode. Here energetic deuterons strike the cathode material and eject positive ions of the cathode material itself, so that the exchange here is 'hot' deuterons for 'cold' metallic ions. Not only is there an energy loss, but we have lost the deuterons which we hoped to fuse together and have replaced them by relatively heavy metallic ions, the nuclei of which have a much higher charge than the deuterons. The coulomb repulsive forces between the heavy nuclei are therefore very much larger so that even higher temperatures would be required to cause them to undergo fusion and, in any event, fusion between such heavy nuclei would not produce appreciable quantities of energy. In short, the plasma has become 'poisoned' by cathode material because magnetic field containment could not be used at the ends but only along the length of the plasma.

This fundamental difficulty was overcome by the British approach to the problem. The ends were effectively eliminated by applying an ingenious suggestion of Sir George Thomson, that is making the conducting gaseous discharge effectively the secondary of a transformer, so that it could be contained in a toroidal discharge vessel. Work on this idea started at Imperial College in 1947 (Ware, 1957) and at the Clarendon Laboratory, Oxford, under Dr. Thonemann who had reached a similar conclusion independently.

For greater security the programmes were transferred to the then A.E.I. Research Laboratory at Aldermaston and A.E.R.E., Harwell, where they culminated in the Sceptre and Zeta experiments respectively in the late fifties and early sixties. The rapidly rising current in the primary windings of the transformer was supplied by discharging a large condenser bank through them and reached tens of thousands of amperes. The primary winding was linked via an iron core to the toroidal discharge vessel in which deuterium at pressures of about 10^{-3} torr rapidly ionised to pass currents in the millions of amperes region, undergoing ohmic heating followed by pinch-force heating in the process. (A typical pinched toroidal discharge apparatus is shown diagrammatically in Fig. 44.)

Early pinched toroidal discharges like early pinched linear

discharges suffered from the fact that the self-generated B_θ magnetic field contained the axial plasma in a condition of unstable equilibrium. Plasma will always try to move from regions of strong magnetic field to regions of weaker magnetic field and a slight displacement or 'kink' in the axial plasma produced conditions in the magnetic field which caused the kink to increase until the plasma came into contact with the walls of the discharge vessel, producing all the undesirable effects associated with the contact of hot plasma with a material surface. Similarly if the plasma became pinched slightly more than normal the surrounding B_θ field became correspondingly stronger causing that portion of the plasma to pinch even further and squeezing the plasma into weaker field regions where it expanded so that the plasma was first divided into a number of broad regions connected by narrow 'necks' and ultimately split up into 'bars' which travelled radially outwards again to hit the walls of the containing vessel. These large-scale or 'gross' instabilities named respectively 'kink' and 'sausage' instabilities are shown in Fig. 43. They were both successfully overcome by providing an additional magnetic field, the so-called B_Z or B_ϕ field parallel to the axis of the discharge vessel. The effect of this field can be understood physically by appreciating that charged particles will spiral about a magnetic field and are consequently 'anchored' to the direction of that field. Since the B_ϕ field was maintained in a stable axial position by external magnetic coils, the deuterons and electrons which made up the plasma were similarly constrained to take up an axial position. The mathematical treatment of the problems is somewhat more sophisticated and leads to a fairly complex relationship between the degree of 'pinch', the self-generated B_θ and the externally applied B_ϕ field which is necessary to avoid these 'gross instabilities'. This is the so-called Suydam criterion

$$\left[\frac{8\pi}{B_\phi{}^2} \left(\frac{dp}{dr} \right) + \frac{r}{4} \left(\frac{1}{\mu} \frac{d\mu}{dr} \right) \geqslant 0 \right]$$

where B_ϕ is the applied field in the axial direction, B_θ the self-generated θ field, dp/dr the rate of pressure drop (pressure gradient) from the centre of the discharge outwards, r the minor radius of the torus and μ the ratio B_θ/rB_ϕ.

Zeta, Sceptre and similar pinched toroidal discharge machines 'Perhapsitron' and 'Alpha' built in America and Russia respectively were designed to satisfy the Suydam criterion and were stable against large-scale instabilities. Although initial performances were encouraging in that fusion was observed, it soon became clear that the fusion processes were predominantly due to directed motion rather than random thermal motion of the deuterons (Jones *et al.*, 1958) and the possibility of a self-sustained 'thermonuclear' fusion reaction was therefore more remote than initially hoped. It is a source of some regret to the author that his only contribution to this field should have been so negative in character.

It was also found that many of the electrons were escaping to the walls of the vessel across the 'containing' magnetic field lines. This energy-loss mechanism limited the temperature attained by the deuterium ions to about one million degrees and the average electron temperatures were observed to be almost an order of magnitude less than this.

Rather more fundamental studies on plasma stability carried out at Princeton in their 'Stellerator' programme suggested that a possible cause of electron loss was the fact that average B_ϕ field strength near the inner circumference of the torus was greater than that near the outer circumference. This was an inevitable consequence of the fact that the inner circumference of the torus was shorter than the outer $(4\pi nI = L_i B\phi_i = L_o B\phi_o$ and since $L_o > L_i$, $B\phi_o < B\phi_i)$* and gave rise to a radial magnetic field gradient across the torus leading to a partial separation of the electrons and deuterons making up the plasma. Large volumes of the plasma were no longer electrically neutral and this could lead to small-scale instabilities of the plasma which resulted in a loss of electrons. The Princeton team suggested that this could be overcome by making the B_ϕ field not a simple axial field but a 'corkscrew' field spiralling within the toroidal vessel in such a way that its average values along the outer and inner circumference of the toroidal discharge were the same. This field was produced by additional external windings (Fig. 45). The so-called 'sheared' fields were used in

---

*L_i and L_o are the inner and outer circumferences of the torus respectively and $B\phi_i$ and $B\phi_o$ the corresponding magnetic fields.

Princeton without any spectacular improvement in the temperatures obtained, possibly because the degree of pinch, and consequently pinch-force heating, was less in the Stellerators than in other toroidal discharge devices. In fairness to the Princeton group, it must be added that their declared objective from the start was to investigate the fundamental problem of the stability of plasmas contained by magnetic fields rather than to produce high temperatures. The Princeton group also contributed a further original idea. This was a magnetic field, shaped in such a way as to strip off the outer layers of plasma as it circulated within the torus. These outer regions were likely to contain a higher percentage of impurity ions due to their interaction with the torus walls and hence to increase the energy loss due to electromagnetic radiation. This cusp-shaped 'divertor' field was to be reintroduced into toroidal discharge investigations some ten years later.

It was left to the Russians to apply the precise plasma-stability work to high-compression plasmas. They realised that if the self-generated B_θ field were in the correct relationship to the externally applied B_ϕ field this would produce the type of sheared or corkscrew field which had been produced in the Stellerators by additional external coils, and to do this higher plasma currents and densities were required (Fig. 46). Their Tokamak pinched-toroidal discharge apparatus employing the principle first became operational in the mid-1960s, and was in fact rather similar to Zeta in general appearance. Initially it achieved only slightly higher temperatures, but it soon became obvious that the electrons were being contained much more effectively in Tokamak than in previous high-density plasma-toroidal discharges. From an analysis of the performance of previous such machines Bohm had shown that the speed of electron loss by diffusion across the confining magnetic field lines ($D_\perp$) appeared to increase with electron temperatures according to the formula

$$D_\perp = \frac{1}{16} \frac{KTe}{eB}$$

where K is a constant (Boltzmann's constant), Te is the electron temperature, e is the electronic charge and B the magnetic field strength. As the temperature increased, therefore, the rate of electron loss increased also, so that if this 'Bohm diffusion law' had proved to be a fundamental law of plasma behaviour it would have been impossible to achieve simultaneously the high temperatures and the high densities required for a fusion reactor. Measurements on the electron density in Tokamak (Fig. 47) showed that these fell off much less rapidly than expected by the Bohm diffusion 'law', and in short the electrons were being contained much more effectively than in Zeta (Golovin, 1969). This was confirmed by measurements of the electron temperatures which were somewhat higher than those of the positive ions rather than almost an order of magnitude less as in Zeta (Table 7).

It is pleasing to note the improvement in the climate of international co-operation which has been evident during the development of the pinched toroidal discharge experiments. Whilst in the mid- and late-1950s Zeta and Sceptre were constructed under conditions of maximum national security, as were similar types of machine in other countries, Tokamak operating in Moscow in 1970 accommodated experimental teams from America, the United Kingdom and other Western European countries working in collaboration. This led to the adoption and extension of the Tokamak principle in other machines which were constructed subsequently in both Western Europe and the United States. These developments will be discussed in some detail later, but first it is necessary to introduce work on a second and fundamentally different type of magnetic-field containing system which has also contributed to the overall fusion picture.

The other main group of containing magnetic-field configurations depends on the so-called 'magnetic bottle' principle. In its simplest form this consists of an axial magnetic field which is stronger at the ends than in the centre. Such a field is easily produced by a long coil or solenoid which has a larger number of turns at the end than in the centre. The physical principles of plasma containment by this method are again relatively simple (Fig. 48); charged deuterons and electrons, which make up the plasma, spiral around the magnetic-field lines and their energy can be considered to be made up of rotational energy associated with the circular motion around the field lines and translational energy associated with their motion along the field lines. As the magnetic field becomes stronger, the fraction of the energy required for the rotational motion increases and that available for translational

TABLE 7. *Performance of Some Fusion Experiments*

Required conditions:

$$T + d \rightarrow {}^4He + n \qquad T_{ion} > 3 \times 10^7 K$$
$$d + d \rightarrow {}^3He + n \qquad T_{ion} > 2 \times 10^8 K$$

$$n\tau > 2 \times 10^{14} \text{ cm}^{-3} \text{ sec}$$

Machine type	T_{ion} (K) Ion temperature	T_e(K) Electron temperature	n (particles/cm³) Density	τ (sec) Containment time	$n\tau$
Linear pinch					
8-metre pinch (Culham, U.K. 1970)	1.5×10^6	2×10^6	2×10^{16}	3×10^{-5}	6×10^{11}
Mirror machines					
Scylla (theta pinch) (U.S.A., 1962)	2×10^7		10^{13}	10^{-5}	10^8
DCX (particle injector) (Oak Ridge, ~1960)($\equiv 3 \times 10^9$K)	300 keV		10^9	10^{-4}	10^5
Phoenix II (neutral particle injector) (Culham, ~1970)	10 keV ($\equiv 10^8$K)	2×10^7	3×10^9	5×10^{-4}	1.5×10^6
Toroidal pinch machines					
Zeta (U.K., ~1960)	$1-2 \times 10^6$	3×10^5	10^{14}	10^{-2}	10^{12}
Stellerators (Princeton, U.S.A., ~1962)	10^6	10^6	10^{12}	10^{-3}	10^9
Tokamak (Moscow, 1966)	2×10^6	3×10^6	5×10^{13}	2×10^{-2}	10^{12}
DITE (neutral particle injector Tokamak) (Culham, ~1976)	1.5×10^7	2×10^7			10^{13}
PLT (neutral particle injector Tokamak) (Princeton, 1978)	4×10^7 (6×10^7 at lower plasma densities)	3.5×10^7	5×10^{13}	3×10^{-2}	1.5×10^{12}

motion is correspondingly reduced. At a certain field value all the energy is required for the rotational motion so that the particles cannot move forward into the regions of still higher magnetic field, but it can move backwards into regions of lower magnetic field. In other words it is reflected by the increasing magnetic field, which acts as a 'magnetic mirror'. By providing such a reflecting field at each end of the region the whole acts as a 'magnetic bottle'.

Such a containing field configuration can be used in association with a variety of heating mechanisms. The simplest in principle of these is the so-called 'θ pinch'. Here the current in the coils providing the magnetic field is suddenly increased. This leads to a sudden increase in the magnetic-field strength. Individual magnetic-field lines can be considered to be compressed towards the axis of the system, and with them the plasma which they contain. As with the toroidal pinch, sudden adiabatic compression leads to a heating of the plasma. Perhaps the most spectacular of the pinch-type experiments was Scylla in which temperatures of 20 million degrees have been recorded. All these devices suffer from the imperfect containment of the plasma at the mirrors. Although the simple physical picture given above is correct in principle, if one of the charged particles near the mirror undergoes a collision with another particle this can give it sufficient velocity in the axial direction to escape. As plasma densities increase towards the required values, the probability of such collisions also increases and the end losses similarly increase, so that in spite of the high temperature recorded, theta pinch-type apparatuses have so far fallen far short of the plasma densities and containment time conditions required. Containment can be improved by using additional magnetic fields provided by four axial conductors named 'Joffe bars'. The resultant magnetic field has a configuration which can best be compared to the shape of an old-fashioned mint humbug, in other words the ends are pinched together and the loss of plasma is correspondingly reduced (Fig. 49).

A magnetic mirror may also be used to contain energetic particles which are projected into this type of field configuration. The principle here is rather different from the plasma-heating mechanisms previously described. Rather than heating relatively 'cold' plasma, deuterons with energies well in excess of that corresponding to 'thermonuclear' fusion temperatures are produced in accelerators and injected into the magnetic bottle. If perfectly contained they would describe spiral paths within the magnetic bottle and by continuously adding particles the required density to achieve a positive energy output as a result of collisions should be reached. This concept differs from that of high-energy deuterons interacting with a gas or solid target in that the energy loss due to ionising collisions in the latter case is appreciably reduced. The complication is that if a deuteron has sufficient energy to enter a magnetic-field configuration of this type it also has sufficient energy to escape. It must therefore be injected in a different form. In the DCX (Direct Current Experiment) at Oak Ridge, Tennessee, ions were produced which consisted of deuterons attached to neutral deuterium atoms. The so-called D_2^+ ions can be considered as being made up of two deuterons sharing a single orbiting electron. After injecting these ions into the magnetic mirror field they were intercepted by a hydrogen arc which caused the dissociation of the D_2^+ ions $D_2^+ = D_1^+ + D_1^0$ into an uncharged deuterium atom and a single charged deuteron (Fig. 50). Being uncharged, the deuterium atom was unaffected by the magnetic field and escaped, whilst the deuteron now had half the initial energy and half the initial mass. Under these conditions it could be retained by the magnetic field, since its radius of curvature depends on its mass and energy ($r = \sqrt{2Em}/Be$) and is consequently halved.

A similar machine 'Ogra' was built in Russia, except that it relied on the collision of D_2^+ ions with impurity gases to produce the necessary dissociation.

A later experiment, Phoenix at Culham (U.K.) in the early 1970s, employed a similar principle except that instead of injecting singly charged deuterium molecular ions, deuterons were first accelerated to a few keV of energy and passed through a jet of gas. Some of the energetic deuterons collected an electron from the gas and proceeded on their path as energetic neutral atoms. In the Phoenix experiment this energetic beam of neutral particles was then passed into a magnetic bottle-type field, and as neutral atoms they were able to enter the field undeflected. Once inside the field, however, the neutral atoms became ionised due to a phenomenon known as Lorentz ionisation. Essentially this is the production of an ionising electric field due to the interaction of a rapidly moving particle with the magnetic field. The deuterons and electrons so formed were trapped in the magnetic field and in principle, as in the DCX experiment, it was hoped to build up a sufficient concentration of trapped energetic deuterons to achieve fusion conditions. When one appreciates that the easily attained deuteron energy of 10 keV if randomised would be equivalent to 10^8K in temperature terms the particle injection principle appears to have much to commend it. However, containment of energetic deuterons by magnetic-mirror fields, even with the added fields from Joffe bars, proved to be difficult and in common with all other magnetic-bottle configurations Phoenix suffered from imperfect particle containment at the mirrors, and the necessary density and containment time conditions were far from being achieved. The experiment is illustrated diagrammatically in Fig. 51.

This and experiments like it did, however, establish the principle of neutral-particle injection into a magnetic field as a heating mechanism, and it seemed an obvious step to use neutral-particle injection as an additional heating mechanism in the efficient containing field which had been established in the Tokamak experiments.

One such experiment, the Divertor and Injector Tokamak Experiment (DITE) carried out at Culham in 1976, incorporated not only the principle of neutral-particle injection as an auxiliary heating mechanism but also the divertor idea which had first been used on the Princeton Stellerators in the early sixties. By using a magnetic divertor field the plasma escaping to the walls was reduced by 50% and the radiation loss by 85%. The use of injected neutral beams increased the ion temperatures reached in Tokamak machines from 2 million degrees to 20 million degrees, and the good electron containment was maintained as shown by the high electron temperatures. The density containment time product also increased to a value of 10^{13}, only a factor of 20 or so less than that required by the Lawson criterion for a 'break-even' fusion system (Table 7). The DITE experiment, which is supported by Euratom countries jointly, was performed using a torus of major diameter about 2.3 metres and minor diameter about 0.5 metre. A plasma current of 200,000 amperes was induced by the relatively stable Tokamak type discharge and the plasma heating was augmented by 300 kW of power in the injected neutral beam (Fig. 52). Comparable experiments have been carried out in France (TFR), Russia (T-11) and the United States (ORMAK).

Energy losses from plasmas tend in general to be proportional to surface area, whilst the energy production is proportional to volume.

As with fission reactors, therefore, the probability of obtaining a sustained reaction increases with size. The performance of DITE and other particle-injector-type Tokamak experiments was sufficiently promising to encourage the planning of yet larger experiments of the same type. One of these PLT (Princeton Large Torus) became operational in the United States in 1976 with a plasma current of 1 million amperes and has recently been reported as achieving fusion temperatures (Table 7). Protracted discussions on the siting of a Joint European Torus (JET) recently resulted in a decision to site this at Culham, where a Euratom team has been carrying out design-study work since late 1973.

The plasma current in JET is designed to be between 3 and 5 million amperes, and the peak power required to produce the required magnetic-field configurations is about 500 MW(e), the output of a medium-sized power station. Up to a further 10 MW of neutral injected beam power will be used. The toroidal vessel will have a major radius of about 3 metres, but will be relatively broad, having a D-shaped cross-section approximately 4 metres high by 2.5 metres wide (Fig. 53). The total weight of the apparatus will approach 3000 tons and the probable cost of the experiment is estimated at 120 million pounds sterling. There is provision for the addition of a divertor at a later stage, and it is hoped that the apparatus will achieve temperatures, plasma densities and containment time approaching those required for a 'break-even' deuterium – tritium fusion reaction.

First operation of this machine is expected in 1983 and progress towards the above objectives is anticipated to involve a three- or four-year programme. Present theoretical predictions indicate that if the outcome of these experiments is successful a working fusion reactor would require between 10 and 30 million amperes of plasma current. This would require corresponding increases in the power input which would be prohibitive if conventional coils were used to provide the necessary magnetic fields, and consequently any economically viable fusion reactor would need to have superconducting coils in order to reduce the power-input requirements.

It would also have to produce its own supplies of tritium to fuel the deuterium – tritium reaction since tritium, being radioactive with a half-life of 12.3 years, does not occur in nature. It can be produced by irradiating lithium with neutrons, both isotopes of lithium producing tritium according to the equations

$$^6\text{Li} + n \rightarrow {}^4\text{He} + {}^3\text{H} \quad \text{and} \quad {}^7\text{Li} + n \rightarrow {}^4\text{He} + {}^3\text{H} + n.$$

The first of these can be produced by slow neutrons but the second requires fast neutrons. Any future thermonuclear reactor, therefore, must produce its own tritium and this could be done by surrounding the fusing plasma, which is a copious source of fast neutrons, with a lithium 'blanket' producing new fuel in much the same way as does the ^{238}U or thorium blanket in a fission breeder reactor. Although it is appreciated that much theoretical and experimental work is yet to be done before a fusion reactor working on the Tokamak principle can be seriously contemplated, there have been preliminary design studies. Figure 54 shows a section through a possible Tokamak fusion reactor. The plasma in the central core would reach temperatures of up to 10^8K, and this would have to be thermally insulated by a vacuum jacket from the lithium 'breeder' blanket. Most of the energy of the fusion reactor is carried away by the fast neutrons and this would be converted into thermal energy in the graphite-moderator region. The heat produced would be extracted using a second liquid-lithium circuit, and the thermalised neutrons escaping from the graphite would be absorbed in a water jacket containing boron salts. Finally a lead shield would be required to attenuate the γ-ray flux. Only outside this could the superconducting coils be placed to provide the Bϕ-containing field. These would be one of the superconducting alloys maintained at about 4K ($-268°\text{F}$).

Thus in relatively close proximity we have almost the lowest temperature known to man and the highest yet to be achieved. Add to this the further complication that we must somehow inject into this central plasma a multi-megawatt beam of neutral particles and probably incorporate a 'divertor', and it will be appreciated that the engineering problems associated with the design of a fusion reactor of the Tokamak type will be very appreciable even when the necessary fusion conditions have been demonstrated on JET or subsequent experiments. The problems of fast-breeder fission-reactor design appear insignificant by comparison.

It is interesting to examine, however, how the whole fusion programme over the past twenty years has contributed to this latest approach. First, the high-density but somewhat unstable discharges of the early Zeta-type experiments and the low-density but more stable Stellerator work were brought together to produce the Tokamak type of discharge. With this slower more stable pinch the pinch-force heating mechanism became less effective and this and ohmic heating were reinforced by neutral particle injection, an idea which had its roots in the rival 'magnetic-bottle' approach. Add to this the divertor concept coming from the early Stellerator work and it will be appreciated that JET is not in any sense a new concept but a synthesis of previous ideas, applied with a greater understanding of the basic theory of plasma stability than was the case when many of them originated twenty years ago.

Whilst the magnetic-field-confinement programme has dominated fusion research over the past two decades, a completely different concept has emerged in recent years. This is based on the heating of a pellet of mixed deuterium and tritium which has been solidified by freezing at low temperatures. The particle density n in such a solid pellet is of the order of 6×10^{22} particles/cm^3 so that to achieve the Lawson criterion $n\tau \geqslant 2 \times 10^{14}$ s cm^{-3} confinement time of less than 10^{-8} second is required as opposed to the few seconds required for normal plasma densities. The problem is clearly how to get sufficient energy into the deuterium – tritium pellet to achieve the fusion temperatures, and the use of laser beams for this purpose was first suggested as long as ten years ago. The concept appears relatively simple, and is illustrated diagrammatically in Fig. 55. A frozen deuterium – tritium pellet is dropped through the laser beam and should absorb enough energy from it to reach the required temperature. As is always the case, however, the simplistic view requires appreciable modification, since the plasma initially formed by heating of the pellet surface is opaque to laser radiation which is then unable to penetrate the interior of the pellet. In order to have any hope of operating successfully the intensity of the laser beam must be carefully programmed to increase in such a way that it induces in the vaporising pellet a compressive shock wave, so that the initial heating is accompanied by a sudden radial compression, produced essentially by the recoil of ions as fusion neutrons escape from the outer regions of the pellet which have been converted to a high-density plasma. In theory the density of the plasma compressed by this implosive shock wave should exceed that of the original solid pellet, making the necessary containment time even shorter, and the energy would be released in a small-scale explosion of duration about 10^{-9} second. This is called inertial confinement.

The explosive power of a tritium – deuterium pellet is in round terms about a million times greater than that of the same weight of T.N.T. so the pellets would have to be extremely small. In theory pellets weighing only one ten-thousandth of a gram if fed into the device at a rate of ten per second would produce a thermal output of 1000 megawatts, which could be converted into about 350 MW(e) of electrical power by conventional means. The actual peak power in the 'explosion', however, would be far in excess of this and its containment in a suitable vessel would be an obvious difficulty.

In addition to the difficulty of correctly programming the laser-beam intensity the overall energy balance of a laser-induced fusion device presents a problem. The usual high-power laser using neodymium glass converts the input power into light with an efficiency of only 0.2% and research is going on to produce a high-power carbon dioxide gas laser which should achieve a more acceptable efficiency of up to 5%. Lasers are also expensive so that the overall cost of an inertial fusion reactor using multiple high-energy lasers could be prohibitive even if the technical problems were overcome. The United States Department of Energy hope, however, to demonstrate a 'break-even' laser-induced fusion device by the mid-eighties. Work on laser-initiated inertial fusion is also going on at the Kurchatov Institute in Moscow and Lisneil in France.

An alternative approach to inertial fusion is to use intense beams of high-energy electrons rather than lasers to heat the deuterium – tritium pellets. This concept has become feasible due to the development of pulsed multi-million-megawatt beams of high-energy (10 MeV) electrons in connection with the defence programme in the United States, and in Russia where similar work is in progress at the Kurchatov Institute. Work in this field is still in a relatively early stage of development and presents many problems in the production and focusing of pulsed million-ampere beams of electrons. Any possible fusion reactor would require up to a hundred such beams focused from all directions on to the pellet. It has recently been proposed that instead of solid pellets the target should consist of hollow shells.

It is still too soon to assess the feasibility of the electron-induced fusion, or indeed of inertial confinement systems generally, but as the size and complexity of the experiments grow it becomes clear that this cannot be regarded as a 'quick and easy' solution to the problem of controlled thermonuclear fusion.

It should be noted that as the fusion research programme generally has developed it has tacitly been accepted that the deuterium – tritium fusion temperature of 3×10^7K represents a more realistic target than does the all-deuterium fusion temperature of 2×10^8K. Reserves of lithium from which tritium is 'bred' therefore became the critical factor in assessing the long-term fission-fuel reserves. Although lithium has not been fully prospected current estimates indicate a world reserve of about 5 million tons of economically recoverable lithium. This would have a fuel value about equal to that of the reasonably rich uranium reserves if used in fast breeders, or to the coal reserves. This is a very considerable fuel resource but not the virtually unlimited one represented by the deuterium reserve.

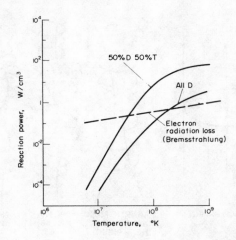

FIG. 41. Energy output and radiation loss vs. temperature for fusion reactions.

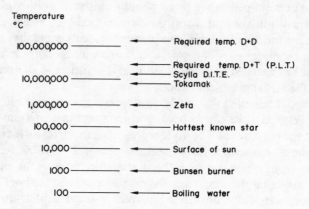

FIG. 42. Some points on the high-temperature scale.

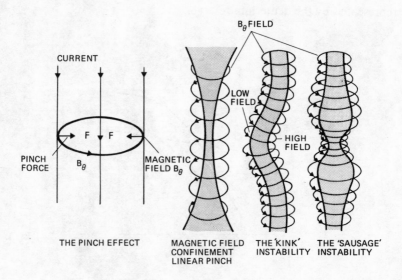

FIG. 43. Containment by B_θ field.

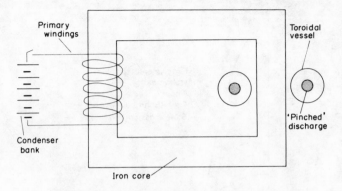

Fig. 44. Section of a pinched toroidal discharge apparatus.

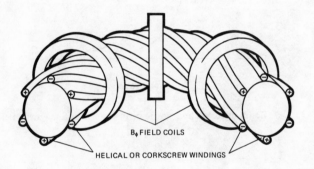

B_ϕ FIELD COILS

HELICAL OR CORKSCREW WINDINGS

Fig. 45. Helical windings, Stellerator.

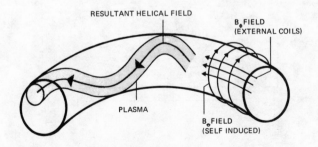

RESULTANT HELICAL FIELD

B_θ FIELD
(EXTERNAL COILS)

PLASMA

B_θ FIELD
(SELF INDUCED)

Fig. 46. Tokamak magnetic-field confinement.

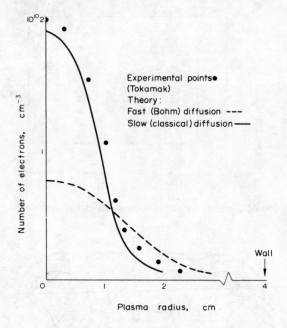

FIG. 47. Bohm diffusion and classic diffusion across a magnetic field.

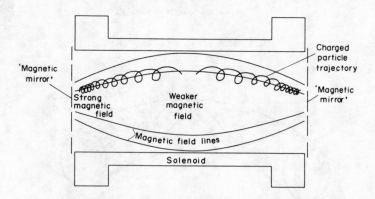

FIG. 48. 'Magnetic bottle' field configuration.

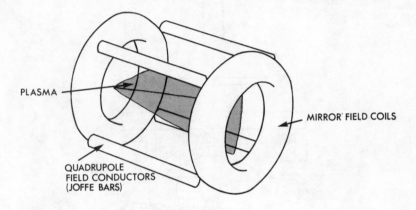

PLASMA

MIRROR' FIELD COILS

QUADRUPOLE
FIELD CONDUCTORS
(JOFFE BARS)

FIG. 49. A modified 'magnetic bottle' configuration with 'Joffe bars'.

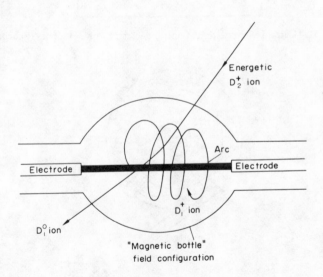

Energetic
D_2^+ ion

Arc

Electrode

Electrode

D_1^+ ion

D_1^0 ion

"Magnetic bottle"
field configuration

FIG. 50. Containment of an injected charged particle after dissociation.

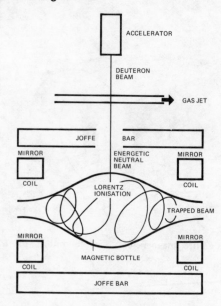

FIG. 51. Neutral beam injection in mirror geometry with Joffe bars (Phoenix).

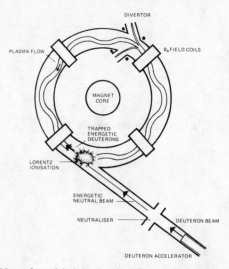

FIG. 52. Neutral particle injection in a Tokamak field configuration.

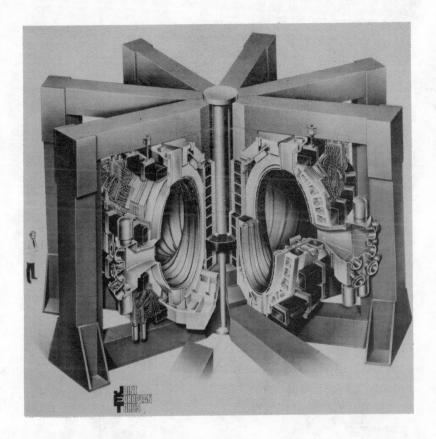

FIG. 53. The proposed JET apparatus. (Reproduced by kind permission of Dr. R. S. Pease, Director of Culham Laboratories.)
The primary windings are accommodated in the outer rectangular sections parallel to the axis of the D-sectioned torus and the B_ϕ field coils are wound round the minor circumference.

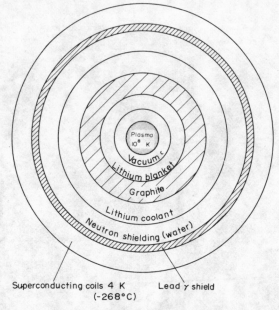

Fig. 54. Section of a possible fusion reactor based on Tokamak.

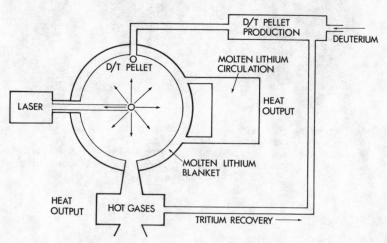

Fig. 55. Schematic diagram of possible fusion reactor based on laser heating of deuterium – tritium pellet.

References

ALLEN, N. L. *et al.* (1958) *Nature*, **181**, 222.

ATKINSON, R. E. and HOUTEMANS, F. G. (1929) *Z. Physik*, **54**, 656.

Culham Laboratory Annual Reports 1970-1977.

FOSTER, C. A., HENDRICKS, C. D. and TURNBULL, R. J. (1975) *Applied Physics Letters*, **28**, no. 10, 580.

GOLOVIN, I. N., DHESTROVSKY, YU. N. and KOSTOMAROV, D. P. (1969) Nuclear Fusion Reaction Conference, Culham.

JONES, W. M., BARNARD, A. C. L., HUNT, S. E. and CHICK, D. R. (1958) *Nature*, **182**, 216.

PEASE, R. S. (1971) International Conference on the Peaceful Uses of Atomic Energy, Geneva.

STEINER, D. (1971) *New Scientist*, **168**.

THONEMANN, P. C. (1956) *Nuclear Power*, Vol. 1, p. 169.

THONEMANN, P. C. *et al.* (1958) *Nature*, **181**, 217.

TUCK, J. L. (1972) *La Recherche*, no. 27, 3, 857.

WARE, A. A. (1957) *Engineering*, 15 Nov. 1957.

YONAS, G. (1978) *Scientific American*, **239**, no. 5, 40.

Summary and Comment

MATERIAL living standards are closely linked to the *per capita* energy consumption and it is clear that if the developing countries are to reach acceptable living standards, the global energy demand must rise rapidly during the next century. This increase cannot be met from our dwindling oil reserves and, although global coal reserves are much greater, their geographical distribution is such that they could not meet the increased demand without excessive transport costs even if political restrictions were overcome. Renewable energy sources are attractive but the most promising of these, solar energy, is intermittent and unpredictable in temperate countries and this will probably limit its application to domestic heating and air conditioning. It should have a larger part to play in the intermediate technology of the hotter developing countries, but as these reach higher levels of sophistication the difficulty of converting solar energy into electricity with reasonable economy will limit its application there as it does in the present industrial countries. Of the other sources wave power may become significant for island communities such as the United Kingdom, but tidal power, geothermal and wind power are likely to remain limited sources of energy in especially favoured localities.

Nuclear power from the fission process remains as the only reasonably assured energy source to meet the probable expansion in demand, and if this is to last beyond the end of the twentieth century it must be based on fuel recycling via breeder reactors, which alone are likely to be able to make economic use of the more dilute but plentiful reserves of uranium. High conversion ratio breeders with a short doubling time are required if the nuclear-generating system is to expand to meet the probable increase in electricity demand.

154

The early development of nuclear power was considerably influenced by the military programme and the military and political situation continues to intrude into the later phases of the development of the civil programme. The main factor inhibiting the development of the highly enriched fast breeder is undoubtedly the fear by the major powers that accumulated $^{239}_{94}\text{Pu}$ or $^{233}_{92}\text{U}$ may become widely available to smaller nations for military use. The present nuclear Non-Proliferation Treaty is an attempt to minimise this danger in that the six nuclear powers agree to provide fuel for nuclear reactors operated by the other nations in return for an agreement that the 'non-nuclear' nations will not themselves process and recycle spent fuel. This appears to the author to be at best a temporary expedient in that the possibility of permanently maintaining two distinct categories of 'have' and 'have-not' nations in nuclear matters is as remote and undesirable as permanently maintaining these same two categories at the general economic level.

The main hope for long-term peace lies in the fact that if and when the smaller nations obtain nuclear weapons their degree of social responsibility or, more to the point, their fear of retaliation will be just as great as that of the major powers. Here it can be asserted that the horror of large-scale nuclear war has served to limit the escalation of 'confrontations' to geographically limited conventional wars in the post-1945 period.

The main danger to this comforting philosophy is the recent development of tactical nuclear weapons, the so-called 'neutron bomb' which could serve to blur the distinction between conventional and nuclear warfare, and thus lead to the escalation of the one into the other.

This danger, and that of nuclear conflict generally, has, however, little to do with the civil nuclear-power programme, whether based on fast breeders or otherwise, since weapons-grade plutonium can be produced much more cheaply and easily by custom-built military reactors. It is, however, one of the reasons which has led to the recent American decision to ban fuel reprocessing and the development of the fast breeder and to discourage it elsewhere.

The present nuclear policy is influenced mainly by short-term economic considerations and the favourable position of the slightly enriched 'burner reactors' compared with fossil-fuelled stations has led to a dramatic increase in their planned numbers in many countries. In retrospect it is regrettable that so much effort has been expended in the development of the various slightly enriched reactor types each with its merits and disadvantages, but rather closely matched economically and having little part to play in the long-term programme. The widespread adoption of the light-water-moderated reactors is even more regrettable since of the slightly enriched reactor types, these are amongst the more demanding in their use of uranium fuel and isotope-separation facilities, and are one of the poorest producers of plutonium. Their possible safety-problem has also attracted opposition to the nuclear-power programme generally. There is a danger that the sudden expansion in the installation of these reactors may jeopardise the long-term future of nuclear power by exhausting the reasonably economic fuel reserves before adequate supplies of plutonium have been accumulated for the fast-breeder programme. One hope is that the large reserves of plutonium at present 'stockpiled' for military purposes will be committed to a fast-breeder programme by the turn of the century. Alternatively, a more slowly developing thermal-reactor programme based on 'near-breeder' reactors is required.

Even despite uncertainties which have been expressed regarding the safety of the light-water-moderated reactors, most investigations agree that the probable cost in human life and environmental damage of a nuclear programme would be very much less than that of an expanded coal-based programme even assuming that the latter were feasible, and the risks associated with some of the renewable energy sources may also be greater than those associated with nuclear power. These predictions are based on calculations of the probability of nuclear accidents on theoretical grounds since over twenty years after the installation of the first nuclear-power stations there has been no instance of death or injury due to reactor malfunction, which is in itself an impressive statistic.

The main environmental problem is the treatment of the radioactive nuclear waste resulting from widespread nuclear-power programmes and here the long-term plans for vitrification and 'burial' of extracted fission products in stable geological formations appears to be a satisfactory solution. The accumulation of much larger volumes of untreated fuel elements which would be the obvious consequence of any extension of the present American ban on fuel reprocessing would present a far worse problem.

The main application of fission reactors is for the production of electricity, and their use for propulsion is clearly limited at present to a fairly narrow range of special applications and of these the author must confess to grave concern at the possible use for rocket propulsion, with all the attendant problems of loss of radioactive material on re-entry or of numbers of reactor cores in permanent orbit. This concern is felt only to a slightly lesser extent to the use of reactors for ship propulsion. Merely on the grounds that the moving objects are more likely to collide than stationary ones it seems possible that, however well reactors are protected, a serious collision could lead to the dissemination of large quantities of radioactive material and the possibility of such a collision must increase if the use of reactors becomes common for merchant shipping. It is at least arguable that the use of nuclear reactors for ship propulsion should be limited to applications where they have a clear-cut operational advantage over conventional engines, preferably in icebreakers rather than military submarines.

Nuclear power may have a less direct but more widespread contribution to make to transport in that an appreciable reduction in electricity costs could, with further development of accumulators, make the 'battery-driven' car more competitive. This would also reduce the widespread pollution problems which the increased use of the internal-combustion engine is producing in most of our large cities.

The present 'battery car' has a range of some 100 miles before recharging and a top speed of about 40 miles per hour. Even this modest performance may not prevent it from playing an effective role in our future transport pattern. It seems clear that on the grounds of overall energy conservation, pollution and traffic congestion, long journeys will have to be made increasingly by electrified railways, and the use of personal transport limited to short journeys between home and rail terminals, peripheral parking areas for large cities, shopping precincts, etc. A limited-range pollution-free vehicle such as the battery car would be ideal for this purpose.

Shortage of food is already a main factor in inhibiting the development of three-quarters of the world's population and, in most cases, this could be alleviated by the provision of adequate supplies of fresh water for irrigation purposes. Even in the developed countries freshwater supplies for industrial and domestic use are rapidly becoming inadequate and extensive desalination programmes appear to be required.

Nuclear heat sources for desalination have the advantage over conventional sources that, not requiring delivery facilities for large volumes of fuel, they can be located in relatively remote areas. Whilst the economics of desalination may not be very attractive, if one considers the fresh-water output alone, it should be appreciated that the value of the mineral by-products, including gold and uranium, should also be taken into account in the overall economic assessment. Since sea water represents the most abundant source of recoverable uranium, a combined desalination and uranium-extraction system using the 'waste heat' from coastal power reactors appears to be a most attractive long-term concept.

Perhaps the only justification for the present short-term rather than the long-term emphasis of the fission programme could be that nuclear fusion can be expected to replace fission reactors before the rich reserves of uranium are exhausted, rendering the conservationist approach to the fission fuel reserves unnecessary.

Compared with the fission process, the fusion process has much to commend it as an energy source. Reserves of deuterium greatly exceed those of uranium, even including the dilute reserves of the latter, although those of lithium, the main source of tritium for the more easily realisable deuterium – tritium fusion process, are more limited. Tritium is also the only radioactive product of the deuterium fusion process and is in no sense a waste product, so that the radioactive waste-disposal problem associated with the fission process is much less in the fusion case. In addition the total amount of radioactive material involved in a possible future fusion reactor would be many orders of magnitude less than in the fission case and the environmental hazard associated with an accident correspondingly less.

Whilst the objectives to be realised in order to achieve a controlled fusion system are now clearly defined, and there has been appreciable progress in the understanding of the basic problems of plasma stability, one is compelled to observe that some thirty years after the first fusion experiments the necessary fusion conditions have as yet to be realised even in a laboratory experiment. The increasing complexity of the required magnetic confining fields appears to indicate that even if thermonuclear fusion is realised experimentally the construction of a reactor will present major engineering problems. Possible inertia confinement systems relying on fusion by laser beams or very high-power electron beams also appear likely to be very large, costly and technically complex.

The problems presented by possible fusion reactors based on present ideas may not be insoluble but it would be foolhardy to rely on fusion as an assured energy source for the foreseeable future.

It is instructive, if a little depressing, to compare the time-tables for the development of the fission and fusion programmes. Whilst only three years were to elapse between the formulation of the concept of a self-sustained fission chain reaction and its realisation in the 'Chicago pile', some thirty years were to elapse before nuclear power became commercially competitive. The frequently quoted maxim that there is a factor of 10 in time and effort between the proof of the feasibility of a new scientific concept in the laboratory and its development into a commercially viable product appears to have been followed closely in the development of the fission programme. The possibility of a similar situation in fusion cannot be excluded.

Apart from these extremely important practical aspects, one of the attractions of both the fission and the fusion processes is that they convert mass into energy with an intrinsic efficiency which is approximately a million times higher than that of the chemical burning process, yet this efficiency is still only one part in a thousand.

It is tempting to speculate whether other processes with an even better mass-to-energy conversion efficiency could be exploited. One such process is the annihilation process in which a positive electron or positron combines with an electron in such a way that both their masses are converted completely into energy in the form of two γ-rays.

On passing through the matter the γ-ray energy ultimately degenerates into heat, so that in principle the process has a 100% efficiency for the conversion of mass into thermal energy. One of the particles, the electron, is in very plentiful supply, but its anti-particle, the positron, is relatively rare. It is produced in the decay of some natural and artificially produced radioactive nuclei and by the pair-production process. This latter process can occur inside the nucleus when it is irradiated by γ-rays, but since the energy required to produce an electron – positron pair is exactly equal to that liberated when the positron annihilates with an electron there does not appear to be any possibility of producing a system with an overall energy release using a combination of the pair production and positron-annihilation processes. Similarly a system which relied on the production of positrons from artificially produced positron emitters appears too complex to have a practical application. The problem of producing large numbers of free positrons with an energy expenditure of appreciably less than the rest mass energy of an electron plus a positron (1.02 MeV) is a basic obstacle to this approach.

Mesons also annihilate with their respective anti-particles, but meson production using high-energy acceleration is prohibitively expensive both in terms of overall energy consumption and cost.

In more general terms, theory indicates that all particles have their negative mass anti-particle, and conversion of a particle into its negative mass anti-particle would liberate energy equal to twice that corresponding to the mass of the initial particle, an apparent efficiency of 200%. Whilst it is easy to dismiss such speculations as theoretical flights of fancy, it is chastening to recall that Rutherford, after contributing so much to our understanding of the atom and its nucleus, dismissed the idea of applying any nuclear reactions for the economic production of energy in almost exactly these terms.

Glossary of Terms

Absolute (Kelvin) temperature scale. Temperature scale referred to absolute zero: $0°C = 273.2$ K, $100°C = 373.2$ K.

A.G.R. Advanced gas-cooled reactor (see page 31).

Alpha. Alpha-particle. Nuclear particle containing two protons and two neutrons tightly bound together.

ALPHA. Russian fusion experiment of the pinched toroidal type.

Atomic mass unit. Unit in which nuclear masses are measured, equal to one-twelfth of the mass of the carbon (^{12}C) atom.

Atomic number (Z). The number of protons in the nucleus, which is equal to the number of electrons in orbit around the nucleus. This is indicated by a subscript before the chemical symbol, $_8O$.

Body burden. The amount of radioactive material within the body. **Maximum permissible body burden.** The amount of radioactive material which can be tolerated within the body without causing measurable damage to health.

Bohm diffusion. Diffusion of electrons from a high-temperature plasma perpendicular to the lines of the containing magnetic field. The rate of diffusion D is given by

$$D_\perp = \frac{1}{16} \frac{KTe}{eB}$$

where K is Boltzmann's constant (1.377×10^{-16} erg deg^{-1}), Te the electron temperature, e the electronic charge and B the strength of the containing magnetic field.

Breeder ratio. See **Conversion ratio.**

Breeder reactions. Reactions by which fertile nuclei ^{238}U or ^{232}Th are converted into fissile nuclei ^{239}Pu and ^{233}U respectively.

$$^{238}_{92}U + n \rightarrow {}^{239}_{92}U \xrightarrow{\beta} {}^{239}_{93}Np \xrightarrow{\beta} {}^{239}_{94}Pu.$$

$$^{232}_{90}Th + n \rightarrow {}^{233}_{90}Th \xrightarrow{\beta} {}^{233}_{91}Pa \xrightarrow{\beta} {}^{233}_{92}U.$$

'Burn up'. The fractions of fissile material undergoing fission in a reactor core before reprocessing is necessary.

B.W.R. Boiling-water reactor (see page 33).

Candu. Heavy-water-moderated and cooled reactor developed mainly by the Canadians (see page 25).

161

Carnot cycle efficiency. The theoretical upper limit to the efficiency of a perfect heat engine working with input and output temperatures T_1 K and T_2 K respectively. This is equal to $(T_1 - T_2)/T_1$. The efficiency with which heat can be converted into electricity is less than this but depends on the output temperature of the reactor working fluid T_2 K and the temperature of the returning fluid T_2 K in approximately the same way.

Chain reaction. A reaction which, once initiated, can continue, e.g. a chain of fission reactions (Figs. 7 and 10) which are initiated by neutrons and in turn produce more neutrons.

Classified worker. A worker who is occupationally exposed to radiation. Such workers are subjected to medical and blood examinations, and the dose to which they are exposed is carefully monitored and must not exceed 5 rems per year.

Cleo. A pinched toroidal-fusion experiment with sheared magnetic field at Culham.

Coffin. A reinforced lead protective container used to transport used fuel elements.

Containment time. The time for which a heated plasma can be held together by the containing magnetic field.

Control rods. Rods of neutron-absorbing material such as cadmium or boron which are used to control the reaction rate in a nuclear reactor.

Conversion ratio. The number of fissile nuclei (e.g. $^{239}_{94}$Pu or $^{233}_{92}$U) which we produce in a reactor for each initial fissile nucleus undergoing fission. If the conversion ratio is in excess of unity it is usually referred to as the breeder ratio.

Criticality. The condition that a nuclear reactor will just sustain a fission chain reaction at constant power.

Sub-critical. A chain reaction in which the rate of fission reactions (and power output) is falling.

Supercritical. A chain reaction in which the rate of fission reactions (and power output) is increasing.

Critical size. The size of a reactor at which a self-sustained fission reactor just becomes possible (see page 21).

Cross-section. The cross-section of a nuclear reaction is a measure of the probability that a nuclear reaction will occur. It is numerically equal to the probability of the reaction occurring when one incident particle strikes a target containing one nucleus per square centimetre. More generally the cross-section, indicated by the symbol σ, is given by

$$\sigma = \frac{\text{number of reactions}}{no \times N \times t}$$

where *no* is the number of incident particles, N the number of target nuclei per cm^3, t the target thickness in cm, so that N_t is the number of target nuclei per square centimetre of target.

The dimensions of σ are those of area, and the probability of a reaction occurring is considered to be proportional to the area of an *imaginary* disc representing the target nucleus and expressed in units of the barns per nucleus, a barn being 10^{-24} cm^2. The actual physical size of the target nucleus is not, however, a major factor in determining the cross-section of a nuclear reaction.

Capture cross-section. Refers to the probability that the incident particle will be captured.

Reaction cross-section. Refers to the probability that the reaction as a whole will take place, that is that the incident particle will be captured *and* that this will be followed by the emission of the particle or quantum as specified in the reaction, e.g. the fission cross-section for the ^{238}U nucleus refers to the combined probability that an incident neutron will be captured by the ^{238}U nucleus *and* that the ^{239}U nucleus so formed will undergo fission.

Curie. A unit of radioactivity equal to 3.7×10^{10} radioactive disintegrations per second.

    Millicurie (mC), one-thousandth of a curie.

    Microcurie (μC), one millionth of a curie.

DCX. Direct current experiment. A fusion experiment at the Oak Ridge National Laboratory, Tennessee, in which singly charged deuterium atoms were injected into a 'magnetic bottle' containing field.

Delayed neutrons. Neutrons which are emitted from certain fission product nuclei after the fission process has occurred.

Deuterium. The heavy isotope of hydrogen, of which the nucleus contains a proton and a neutron.

DITE. Divertor injected Tokamak experiment, Culham, 1976 (Fig. 52).

Dose rate. The rate at which the body is exposed to radiation. The maximum permissible level (M.P.L.) of the dose rate for unselected members of the population is 0.25 millirem per hour.

Einstein's mass – energy equivalence. Mass is a form of energy and a body at rest has energy equal to $E = m_0c^2$ where m_0 is the mass of the body and c is the velocity of light. In chemical and nuclear reactions a fraction of the initial mass may be converted into kinetic energy which usually degenerates into heat energy, or energy may be absorbed so that the final mass is greater than the initial mass. These are exothermic and endothermic reactions respectively. Typically about one part in a thousand of the total mass present is converted to energy in the case of nuclear reactions and one part in a thousand million in the case of chemical reactions.

Electron volt (eV). A unit of energy used in nuclear physics. It is equal to the kinetic energy which a singly charged particle (e.g. an electron or a proton) gains in falling through a potential difference of one volt. In practice the energy of uncharged particles (neutrons) and quanta (X- and γ-rays) is also expressed in terms of electron volts, or multiples of this unit. The energy of a slow neutron due to thermal motion at normal temperatures is about one-fortieth of an electron volt.

    Kilo-electron volt (keV). One thousand electron volts.

    Million electron volts (MeV). One million electron volts. A proton or neutron of energy 1 MeV travels at a velocity of about 25 million miles per hour.

Endothermic or endoergic reaction. A reaction which absorbs kinetic energy (see **Einstein's mass – energy equivalence**).

Enrichment. To increase the percentage of fissile material in a reactor core above the 0.7% of ^{235}U found in natural uranium. This may be done by adding either ^{235}U from isotope-separation plant or ^{239}Pu or ^{233}U produced by the breeder reactions in previous reactors (see page 59).

Exothermic or exoergic reaction. A reaction in which kinetic energy is produced (see **Einstein's mass – energy equivalence**).

Farmer curve. A curve relating the possible release of radioactive material in a reactor accident to the theoretical possibility of such an accident occurring (Fig. 22).

F.B.R. Fast-breeder reactor (Figs. 18 and 19).

Fertile nucleus (material). A nucleus which on capturing a neutron is converted into a fissile nucleus (see **Breeder reactions**).

Fissile nucleus (material). A nucleus which undergoes fission on capturing a neutron. ^{235}U is the only naturally occurring fissile nucleus for slow neutrons and two, ^{239}Pu and ^{233}U, can be produced artificially by the breeder reactions.

Fission product nuclei. The nuclei which are produced as a result of the fission process (Fig. 5, page 17).

Fleming's left-hand rule. Rule relating the directions of the electric current, magnetic field and resultant mechanical force. When the thumb, index and middle finger of the left hand are held mutually at right angles, if the forefinger is placed in the direction of the magnetic field lines and the middle finger in the direction of the current, the thumb will indicate the direction of the force.

Fused salt reactor. A reactor using fused salts of fissile material as the fuel and primary heat transfer medium (pages 35, 39 and Fig. 20).

Half-life $T_{1/2}$. The time required for the radioactivity of a sample to fall to half its initial value.

Heavy water. Water in which normal hydrogen is replaced by its heavy isotope, deuterium (chemical formula D_2O).

Heterogeneous reactor. A reactor design in which the fuel and moderator are separate.

Homogeneous reactor. A reactor design in which the fuel and moderating material are intimately mixed.

H.T.R. High-temperature reactor (page 32).

I.C.R.P. (International Commission on Radiological Protection). Commission responsible for recommending safe procedures for handling and transporting radioactive material, safe use of radiation-producing machinery, maximum permissible levels of radiation, radioactive contamination, etc.

Ionisation. The removal of an electron from a neutral atom to produce a positively charged particle or ion.

Isotones. Nuclei with the same number of neutrons but different numbers of protons. Since the electron shell configurations are therefore different, isotones have different chemical properties.

Isotopes. Nuclei with the same number of protons but different numbers of neutrons. Since they have the same electron-shell configurations they are chemically identical.

Isotope separation. A system for separating isotopes. This cannot be done chemically and present methods include electromagnetic separation, and separation by gaseous diffusion and by gas centrifuge (page 59).

JET. Joint European Torus (Fig. 53).

Kelvin scale. Temperatures on the Kelvin or Absolute scale are measured with reference to Absolute zero temperature ($-273.2°C$).

Kinetic energy. Energy of motion equal to $\frac{1}{2}mv^2$ where m is the mass and v the velocity of the particle.

Lawson criterion. Criterion specifying the relation between the containment time and the plasma density of a fusion device in order to achieve an overall kinetic energy gain (assuming that an adequate temperature has also been achieved) (see page 130).

Lethal dose (median). That dose of radiation which, if received over the whole body in a short time, will kill 50% of the people so exposed. This is taken as 400 rems.

Magnetic bottle. A confining magnetic-field configuration using a magnetic mirror at each end, used in fusion research (see page 136).

Magnetic mirror. A region of increased magnetic field used to reflect plasma in fusion research (see page 138).

Magnox reactor. Natural-uranium graphite-moderated CO_2-cooled reactor (see Fig. 11 and page 24).

Mass number (A). The total number of nucleons (neutrons plus protons) in a nucleus. This is indicated by a superscript before the chemical symbol, ^{16}O.

M.P.C. Maximum permissible concentration (of radioactive material). The highest level of contamination by radioactive material which can be tolerated throughout a normal life span without obvious injury to health.

M.P.L. Maximum permissible level (of radiation). See **Dose rate**.

M.S.T.B.R. Molten-salt thorium breeder reactor (Fig. 20).

Multiplication factor (K) (also known as reproduction constant). The ratio of the number of fission processes taking place in one 'generation' of a fission chain reaction to the number occurring in the previous generation.

Ogra. A Russian fusion experiment using injected singly changed deuterium atoms in a magnetic-bottle-confining field (page 139).

Ohmic heating. The heating of a plasma produced by passing a current through it, equal to I^2R where I is the current and R the plasma resistance (see page 130).

Pebble-bed. Pebble-bed H.T.R. (Fig. 14).

Phénix. French fast-breeder reactor 250 MW(e). Super Phénix planned, 1300 MW(e).

Phoenix. Neutral beam injected fusion experiment (Culham) (Fig. 51).

Pinch-force heating. Heating of a plasma due to adiabatic compression caused by the interaction of the current flowing through the plasma and the self-generated magnetic field (see Fig. 43 and page 131).

Poison (reactor). Fission product nuclei which capture neutrons and reduce the reactivity of a reactor during operation (see page 47).

Prompt neutrons. Neutrons emitted at fission.

P.W.R. Pressurised-water reactor (see Fig. 15 and page 33).

Q. A unit of energy equal to 10^{18} British thermal units or approximately 3×10^{14} kWh. It is also equal to the energy produced by burning 46,500 million tons of coal.

    Q value (of a nuclear reaction). The amount of mass converted into kinetic energy or kinetic energy into mass as a result of a nuclear reaction. If mass is converted into energy the Q value is positive and the reaction is exothermic or exoergic. If energy is converted into mass the Q value is negative and the reaction is endothermic or endoergic (see page 11).

Rad. A unit of absorbed X- or γ-ray dose. That amount of X- or γ-radiation which produces 100 ergs of absorbed energy due to ionisation in 1 gram of any material.

Millirad (mR). One-thousandth of a rad.

Radioactivity. The decay of unstable nuclei by the emission of α- or β-particles or γ-rays.

Reactivity (ΔK). The fraction by which the multiplication factor differs from unity.

$$\Delta K = \frac{K - 1}{K}$$

Excess reactivity. When the reactivity is positive ($K > 1$) it is often referred to as excess reactivity.

Rem. That amount of other types of radiation which produces the same biological damage as one rad of X- or γ-radiation.

Millirem (mrem). One-thousandth of a rem.

Research reactor. A reactor used for the purposes of research rather than the production of electricity. Research reactors may either primarily be used as sources of neutrons for irradiation purposes or to investigate specific problems associated with power reactors.

Rest mass energy. See **Einstein's mass – energy equivalence.**

Salter's Ducks. Floats for conversion of wave power (Fig. 40).

Scylla. An American fusion experiment of the fast-pinch type (see page 138).

S.G.H.W. Steam generating heavy-water reactor (see Fig. 17 and page 34).

Shuffling. Interchanging fuel rods within the core of a reactor to equalise 'burn-up'.

Shutdown (safety) rods. Rods of neutron-absorbing material (usually cadmium or boron or compounds of these) which are normally held outside the operating reactor core, but can be inserted to shut the reactor down in case of emergency. Spheres of neutron-absorbing material replace the shutdown rods in some designs and neutron-absorbing salts may also be used in liquid-moderated reactors.

Stellerators. Fusion experiments of the pinched-toroidal type but employing 'sheared' or 'corkscrew' axial magnetic fields for improved stability, originated in Princeton, U.S.A.

Swimming-pool reactor. A light-water-moderated and cooled research reactor usually having a small highly enriched mobile core. Used primarily as a source of neutrons for research irradiations.

Thermonuclear reaction. A fusion reaction produced by the thermal agitation of the fusing deuterium or tritium nuclei. Such a reaction being produced by heat and in turn producing more heat could be self-sustaining (see page 129).

Tokamak. The Russian pinched-toroidal discharge experiment using sheared magnetic fields and a high compression ratio (page 135).

Torr. A unit of pressure equal to that exerted by a column of mercury 1 mm high. Atmospheric pressure is about 760 torr.

Wigner release. Release of stored energy from irradiated graphite (see page 52).

Zeta. A pinched-toroidal discharge apparatus first operated at Harwell about 1960.